Clinical Physiology and Pharmacology

Clinical Physiology and Pharmacology

The Essentials

Farideh Javid

Division of Pharmacy and Pharmaceutical Sciences, School of Applied Sciences, University of Huddersfield, UK

and

Janice McCurrie

School of Pharmacy, University of Bradford, UK

A John Wiley & Sons, Ltd., Publication

This edition first published 2008

Wiley-Blackwell is an imprint of John Wiley & Sons, formed by the merger of Wiley's global Scientific, Technical and Medical business with Blackwell Publishing.

Registered office: John Wiley & Sons Ltd, The Atrium, Southern Gate, Chichester, West Sussex, PO19 8SQ, UK

Other Editorial Offices:
9600 Garsington Road, Oxford, OX4 2DQ, UK
111 River Street, Hoboken, NJ 07030-5774, USA

For details of our global editorial offices, for customer services and for information about how to apply for permission to reuse the copyright material in this book please see our website at www.wiley.com/wiley-blackwell

ISBN 978-0-470-51852-6 (HB) 978-0-470-51853-3 (PB)

A catalogue record for this book is available from the British Library.

Typeset in 10.5/12.5 Minion by Laserwords Private Limited, Chennai, India
Printed and bound in Singapore by Markono Print Media Pte Ltd

First printing 2008

Contents

I've read this book from cover to cover but, I can't remember anything about it today
Me too! what we need is a set of clinical case studies to help us put these facts into our deep memory

Preface

Introduction

Physiology plays a major role in the scientific foundation of medicine and other subjects related to human health and physical performance. Pharmacology is the science which deals with the effects of drugs on living systems and their use in the treatment of disease. This book is designed to enhance students' understanding of physiology and pharmacology via a series of case studies involving human disease and its treatment.

Traditional university teaching methods focus on informing students in terms of physiological and pharmacological theory. This approach, although often extremely efficient and effective, may leave students in a position of remembering the facts and understanding the mechanisms but not necessarily being able to apply their knowledge to real-life situations. The latter ability is a skill which requires time and experience to develop and its acquisition is a key goal in vocational programmes, such as those associated with the training of doctors, pharmacists and other health care professionals. In our own teaching we have found that one very effective means of acquiring this all-important skill is via the use of clinical case studies. The case studies bring basic physiology and pharmacology to life, allowing students to examine ways in which the disruption of homeostatic mechanisms results in patients presenting with specific signs and symptoms. Case studies also enable students to understand how these signs and symptoms can facilitate diagnosis, and this is augmented as the students gain understanding of ways in which pharmacological intervention can be used to treat disruptions in homeostasis.

This book consists of a series of chapters containing case studies organized by major organ system; the book also contains answers to all the questions. There are very few texts available that use clinically relevant case studies to facilitate a student-centred learning approach. This book is designed to fill that niche. This type of student-centred learning not only brings theoretical subjects to life but also promotes deep learning, reflection and enhances analytical skills. We hope you enjoy working through these cases and would be happy to receive your comments on this book to inform future editions.

Aims of the Book

The case studies and the questions which follow will aid your understanding of many types of biological and clinical factors. They are intended to help you prepare for problems associated with clinical physiology and pharmacology that you may meet both in formal examinations and in future professional practice. The case studies presented cover a wide range of psychological, neurological, endocrine, cardiovascular, respiratory, renal, gastrointestinal and reproductive disorders, their symptoms, complications and usual treatment along with the actions, dosage and uses of some widely used drugs. The key points for each case study, which can be found in the Answers section will aid your revision of the major factors associated with each disease or condition.

These case studies provide a practical illustration of common disease states, together with their treatment; the explanations given will help you to relate these conditions to knowledge gained from your lecture courses.

Learning Outcomes

After successfully completing each case, you should be able to:

- understand and describe the signs and symptoms of the disorder in question and its underlying pathophysiology;
- understand and describe the pharmacology of agents currently used in the treatment of the disorder studied;
- appreciate some of the key issues in determining appropriate medication;
- continue to develop your problem-solving skills.

Using This Book

Clinical Physiology and Pharmacology is written primarily for undergraduate students studying modules in physiology and pharmacology as part of a degree in science, pharmacy, preclinical medicine or other health-related courses.

One of the challenges in studying physiology and pharmacology is the very large number of facts and ideas that must be remembered; this factual load can seem daunting. To understand how drugs produce their therapeutic effect, it is essential to have knowledge and understanding of both the physiological mechanisms which underpin pharmacology and the mechanisms of action of drugs currently being used. In addition the innovations of the pharmaceutical industry ensure that the extensive list of therapeutic drugs to be considered continues to increase each year.

Isolated facts, physiological mechanisms, drug names and actions can sometimes be remembered for only a comparatively short time. However, this process of memorizing and understanding facts represents only the first step in your learning.

The next vital stage is to develop your ability to interpret, analyse and use this information in order to solve problems and formulate solutions. Using what you have remembered from your physiology and pharmacology studies to interpret the cases presented in this book will help to move factual knowledge from your superficial memory into deep-memory stores, illustrate the clinical application of this basic knowledge, assist you in revising many important topics and improve both your skills and confidence in problem-solving. Since the information is placed in a realistic setting, your recall of key facts and concepts in physiology and pharmacology will be enhanced.

We hope that using this book will also prove to be a useful step towards applying these skills during your future professional life.

The Case Studies

The case studies are presented as short scenarios with interlinked questions that will both challenge your understanding and lead you through the major learning outcomes of the case as it unfolds.

The *learning outcomes* to be achieved are clearly stated at the beginning of each case study and will focus your attention on the most important facts, topics, mechanisms and concepts to be addressed as you work through it.

Although each case study presents a unique scenario, some important physiological mechanisms and pharmacological agents are involved in more than one of the scenarios. This will give you the opportunity to rehearse knowledge already gained from a previous case study to answer a question directly and enable you to revise any aspects that were not previously clear. The overlap between cases will also help to emphasize that some signs and symptoms are common to several different conditions and that care must be taken to consider all the factors presented before formulating your answers or coming to a conclusion about the case study.

Key points are provided for each case and are intended both as a short summary of the essential points and as a focus for revision. They can be used to preview or review the case content. Important points should then be easier to remember in the future, especially when, by association, you can recall them in an appropriate clinical context.

The *glossary* collects simple definitions of the most important terms into a single location for easy reference.

The *index* lists the number of the case in which the key terms, conditions and drugs are discussed.

The drug doses stated in this book were checked at the time of writing but may now have changed due to revision or updating of treatment regimes. Current dosage recommendations are available in the up-to-date British National Formulary or any other Formulary.

Farideh Javid
Janice McCurrie

The Case Studies

CASE STUDIES

1
Psychological disorders

CASE STUDY 1 A mother's loss

Learning outcomes

On completion of the following case study, you will be able to:

- describe the signs and symptoms associated with this disorder;
- describe the underlying pathophysiology of the disorder presented;
- outline pharmacological approaches to the management of the symptoms;
- explain how drugs may cause their clinical benefits and side effects;
- outline the mechanism of action of amitriptyline hydrochloride;
- explain the advantages of using SSRIs (selective serotonin re-uptake inhibitors) compared to tricyclic antidepressants and MOIs (monoamine oxidase inhibitors).

Clinical Physiology and Pharmacology Farideh Javid and Janice McCurrie

Part 1

It has been nearly five months since 45-year-old Mrs Ford lost her only son. He was 12 years old and was killed in a car accident while playing with his friends. She has been feeling very down since it happened and has an overall feeling of utter hopelessness. She is unable to feel happiness, has difficulty sleeping and her appetite is greatly reduced. Mrs Ford used to enjoy socializing with her friends; however, now she has lost interest. She had been planning to redecorate the house, but since the loss of her son she cannot be bothered. She does not want to cook and when hungry does not feel like eating. She feels that life has no meaning without her son and wishes to join him very soon. Fortunately, Mrs Ford's sister visited her recently and was so worried about her condition that she convinced her to see a doctor. After visiting her family doctor, Mrs Ford was prescribed amitriptyline hydrochloride. The doctor advised her to take this medication at night.

Q1 What is the likely diagnosis of Mrs Ford's symptoms?

Q2 List the symptoms of depression.

Q3 Which of Mrs Ford's symptoms are consistent with the profile of depression?

Q4 Comment on the pathophysiology of this condition.

Q5 What treatments are available for depression?

Q6 Name three categories of drug currently used to treat patients with depression and comment on their mechanisms of action.

Q7 To which category of drug does amitriptyline hydrochloride belong?

Q8 What is the recommended adult dose for amitriptyline hydrochloride? Why was Mrs Ford advised to take the medication at night?

Q9 What are the possible side effects associated with the use of amitriptyline hydrochloride?

Part 2

A week later Mrs Ford made another appointment with the doctor, complaining that the prescribed medication was not effective.

Q10 Can you suggest an explanation for the amitriptyline hydrochloride being ineffective? Does Mrs Ford need a different medication?

Part 3

Three weeks later Mrs Ford visited her doctor again. She reported that her mood had improved and that she felt better than before; however, she complained about having a dry mouth and blurred vision. An alternative drug was prescribed, which proved to be more suitable for Mrs Ford.

Q11 Suggest an alternative drug which is likely to be more suitable for Mrs Ford.

Q12 Outline the advantages of using SSRIs compared to tricyclic antidepressants. Your answer should include an example of an SSRI and its recommended daily dose.

Q13 Name the main side effects associated with the use of SSRIs.

Q14 This patient was not prescribed an MOI. Comment on the disadvantages of using MOIs in the treatment of depression.

CASE STUDY 2 A dangerous father?

Learning outcomes

On completion of the following case study, you will be able to:

- present an overview of mania, its aetiology and associated symptoms;
- outline a possible connection between the use of antidepressants and the development of mania;
- explain therapeutic approaches to managing the symptoms;
- explain the limitations associated with the use of lithium.

Fifty-six-year-old Mr Watson was taken to his doctor by his daughter, who described her dad's condition as being critical and possibly dangerous. She explained that her dad was extremely overexcitable, irritable and angry most of the time; he had developed the delusion that he was in possession of special powers and was showing inappropriate elation. She also mentioned that he had been taking antidepressants for a while, following her mother's death one year earlier.

The doctor made a diagnosis and prescribed lithium, advising Mr Watson to stop taking his antidepressant medication and also not to take non-steroidal anti-inflammatory drugs in combination with his new medication.

Q1 What is your diagnosis of Mr Watson's condition?

Q2 What are the symptoms of mania?

Q3 Outline the underlying pathophysiology of mania.

Q4 Is there a relationship between the development of mania and the use of antidepressants?

Q5 When lithium therapy is initiated, what is the recommended daily dose?

Q6 Describe the mechanism of action of lithium.

Q7 Comment on the side effects associated with the use of lithium.

Q8 Why was Mr Watson advised not to take non-steroidal anti-inflammatory drugs in combination with lithium? Are any other medications contraindicated for patients taking lithium?

Q9 Identify alternative drugs which can be used for patients with mania.

CASE STUDY 3 Continual concerns for Mr Watson

Learning outcomes

On completion of the following case study, you will be able to:

- present an overview of manic depressive disorder (bipolar affective disorder) and the associated symptoms;
- describe its pathophysiology and pharmacological approaches to managing the symptoms of manic depressive disorder;
- explain the clinical benefits and side effects of the drugs used.

Mr Watson has now been on medication to treat his mania for the past year. Recently, his daughter consulted their doctor again, expressing concerns about her father's condition. She explained that her father is now experiencing two opposing mood states: these range from depression to periods when he becomes agitated, extremely talkative and does not want to go to sleep. His mood then appears elevated and euphoric and these irritable moods can last for weeks. On further questioning by the doctor, it became clear that her paternal grandfather had also suffered similar mood swings.

Q1 What is the likely diagnosis for Mr Watson?

Q2 Comment on the pathophysiology of mood swings in manic depressive disorder.

Q3 What is the recommended medication for patients with manic depressive disorder?

Q4 What is the recommended dose of lithium for long-term therapy? Are any special precautions necessary when patients are treated with this agent?

Q5 Name an alternative medication (including the daily dose) suitable to treat manic depressive illness.

Q6 Is the fact that Mr Watson's father also suffered from mood swings significant?

Q7 What advice should be given to patients with manic depressive illness?

CASE STUDY 4 A scary presentation

Learning outcomes

On completion of the following case study, you will be able to:

- describe anxiety and the associated neurotransmitters;
- describe symptoms of anxiety, including somatic and psychological symptoms;
- describe its pathophysiology;
- outline the mechanisms of action of common anxiolytic agents;
- explain the mechanism of action of benzodiazepines;
- explain the connection between anxiety, phobia and panic disorder.

Jo had been asked to give a seminar as part of her final-year project. She was anxious to perform well and spent one month preparing for the presentation. During the preparation period, she was irritable, restless and had difficulty in concentrating; she also complained of diarrhoea. Jo asked some of her friends if they would listen to her practise, prior to the final presentation. But as the day of the practise presentation approached, Jo became very tense, pale and sweaty. She felt increasingly apprehensive and uncomfortable, was unable to talk properly as her mouth was dry and she was very aware that her heart was beating rapidly (tachycardia). She visited her doctor to ask for help as she felt unable to carry on with her normal duties in life.

Q1 What is the likely diagnosis of Jo's symptoms?

Q2 List the symptoms of anxiety.

Q3 Outline the somatic and psychological symptoms evident in this case.

Q4 Which neurotransmitters are mainly associated with anxiety?

Q5 What is the explanation for Jo's tachycardia (increase in the heart rate)?

Q6 Which other conditions could be confused with anxiety?

Q7 What could the doctor prescribe for Jo?

Q8 What are anxiolytics? Your answer should cover the major subdivisions of this class of drug.

Q9 By giving an example of a benzodiazepine, explain the mechanism of action of the named agent in anxiety.

Q10 What are the main concerns associated with the use of benzodiazepines?

Q11 Explain the mechanism of action and usual daily dosage of an anxiolytic agent which does not belong to the benzodiazepine class.

Q12 Can anxiety develop into a phobic state and/or a panic disorder?

CASE STUDY 5 Fussy Jane

Learning outcomes

On completion of the following case study, you will be able to:

- present an overview of obsessive–compulsive disorder and the associated symptoms;
- describe its pathophysiology;
- explain the pharmacological approaches to managing the symptoms of this condition.

Finally, after checking the luggage several times, Jane and her husband managed to get out of the house in time to go to the airport for their holiday abroad. On their way to the airport, Jane asked her husband if they could go back and check the front door once more. She was not sure that the door was properly locked. Her husband reminded Jane that she had checked the door twice before they left. However, Jane did not take 'no' for an answer and insisted on going back to the house. In the past year, her husband had become increasingly aware of Jane's odd behaviours and was fed up with her unnecessary checking of everything several times. Even in the kitchen she would re-wash the crockery, clean all the surfaces several times and repeatedly wash her hands. After returning from their holiday, her husband persuaded Jane to visit their doctor.

Q1 What is the likely diagnosis of Jane's symptoms?

Q2 What are the characteristics of obsessive–compulsive disorder?

Q3 What is the underlying pathophysiology of this condition?

Q4 (A) Name three drugs that can be prescribed for patients with obsessive–compulsive disorder. (B) Comment on the mechanism of action of the drugs you have mentioned in part A.

Q5 Are any other treatments suitable for this condition?

CASE STUDY 6 David's withdrawal

Learning outcomes

On completion of the following case study, you will be able to:

- describe schizophrenia and its associated positive and negative symptoms;
- describe the causative factors and associated neurotransmitters;
- explain the pharmacological approaches to managing the symptoms;
- explain how neuroleptic drugs may produce their clinical benefits and side effects;
- outline the benefits of using haloperidol in schizophrenic patients.

Emma made an appointment for her 27-year-old brother, David, to visit his doctor and persuaded him to keep the appointment. She has been very concerned about his recent behaviour and thoughts. David claims to be able to see and talk to his mum, who died 10 years ago. Recently, he has avoided visits to his local football club and he no longer mixes with his friends. Sometimes he talks very slowly and quietly but on some occasions he is very loud and violent in speech. David has not previously been a religious man, but recently he keeps talking about God. He appears to think that God is talking to him, asking him to perform certain tasks. David was initially very reluctant to talk to the doctor but eventually revealed that he thought his sister was trying to poison him, so he had stopped eating at home. His doctor made a diagnosis and prescribed haloperidol.

Q1 What is the likely diagnosis of David's symptoms?

Q2 What are the positive symptoms of schizophrenia?

Q3 What are the negative symptoms of schizophrenia?

Q4 Can both positive and negative symptoms occur together?

Q5 Identify the positive and negative symptoms presented in this case.

Q6 Are all David's symptoms consistent with the profile of schizophrenia?

Q7 What other conditions could be confused with schizophrenia and should be eliminated before a final diagnosis is made?

Q8 What is the main neurotransmitter associated with schizophrenia?

Q9 What are the possible causes of schizophrenia?

Q10 To what category of drugs does haloperidol belong? Comment on the mechanism of action of haloperidol.

Q11 Name other neuroleptic drugs you know of and comment on the problems associated with neuroleptic therapy.

CASE STUDY 7 Forgetful mum

Learning outcomes

On completion of the following case study, you will be able to:

- describe Alzheimer's disease and its associated symptoms;
- describe its pathophysiology and associated neurotransmitters;
- outline pharmacological approaches in managing its symptoms;
- describe the benefits in the use of anticholinesterase inhibitors in managing the symptoms and their associated drawbacks.

Robina was very worried about her mum who is in her early sixties. It was the second time that her mum had forgotten to pick up her granddaughter from school. She noticed that her mum was becoming increasingly absent-minded and that, although Robina repeated everything that her mum needed to do on a daily basis, she still forgot to do it. This reminded Robina of her grandmother, as she was also very absent-minded and needed help in managing her daily routine tasks. Her mum had previously revealed that there was some history of being absent-minded in the family and, as her mum's condition was getting worse, Robina made an appointment for her to see their doctor. The doctor made a diagnosis and referred the patient to a local specialist clinic, where donepezil was prescribed.

Q1 What is the likely diagnosis for Robina's mum?

Q2 What is Alzheimer's disease?

Q3 Comment on its pathophysiology.

Q4 Which neurotransmitter is mainly associated with Alzheimer's disease?

Q5 To which category of drugs does donepezil belong?

Q6 Comment on the mechanism of action of cholinesterase inhibitors.

Q7 What are the adverse effects associated with cholinesterase inhibitors?

Q8 Are other drugs effective in Alzheimer's disease?

CASE STUDY 8 Disruptive John

Learning outcomes

On completion of the following case study, you will be able to:

- present an overview of attention deficit hyperactivity disorder (ADHD) and its associated symptoms;
- describe the pathophysiology of ADHD and the pharmacological approaches to managing its symptoms;
- outline the difference between attention deficit disorder (ADD) and ADHD;
- explain the mechanism of action of methylphenidate and its side effects.

Mrs Jackson finished a meeting with the headmaster of her son's school. This was the third meeting since the start of this academic year. Her son John, who is only six years old, started school two months ago. Whilst his teachers could understand some hyperactivity in a six-year-old child, they expressed serious concerns regarding John's disruptive behaviour in the class. They found that John was having difficulty in focusing and was unable to remain in his seat, even for a short period. Mrs Jackson was very upset about all this and decided that John should see the family doctor. A diagnosis was made and the drug, methylphenidate, was prescribed.

Q1 What is the likely diagnosis of John's disruptive behaviour?

Q2 What is ADHD? Comment on the subtypes of ADHD and its symptoms.

Q3 Comment on the pathophysiology of ADHD.

Q4 What is the difference between ADHD and ADD?

Q5 To which category of drugs does methylphenidate belong and what dose is recommended?

Q6 What is the mechanism of action of methylphenidate?

Q7 What are the side effects associated with the use of methylphenidate?

Q8 Is ADHD a lifelong condition?

2
Neurological disorders

CASE STUDY 9 Mrs Smith's tremor

Learning outcomes

On completion of the following case study, you will be able to:

- describe Parkinson's disease and its associated symptoms;
- explain the actions of neurotransmitters involved in its pathophysiology;
- outline therapeutic approaches in managing the symptoms;
- compare the effects of dopamine-related drugs and anticholinergic therapies.

Mrs Smith, a retired maths lecturer, is 69 years old. She has consulted her family doctor complaining that she feels very stiff and has developed tremor in her limbs, especially in her hands. She also reported having difficulties getting up and down the stairs at home. She mentioned that her mother remains very fit; however, her father, who died at the age of 70, developed similar symptoms when he was 65 years old. Her doctor made a provisional diagnosis and referred her to a specialist clinic. The consultant prescribed levodopa (L-dopa) plus carbidopa. After finding out the

Clinical Physiology and Pharmacology Farideh Javid and Janice McCurrie

diagnosis of her problems and the prescribed medication, Mrs Smith argued that her friend, with a similar diagnosis, had been prescribed amantadine.

Q1 What is the likely diagnosis of Mrs Smith's symptoms?

Q2 What are the symptoms of Parkinson's disease? Could it be hereditary?

Q3 Comment on the pathophysiology of Parkinson's disease.

Q4 Comment on the pharmacological management of this condition. What classes of drug are available for patients with Parkinson's disease?

Q5 Present the rationale for prescribing carbidopa in combination with L-dopa in Parkinson's disease.

Q6 Why was dopamine not prescribed?

Q7 Comment on the action of amantadine and explain the possible reason for its use.

Q8 Why may antimuscarinic drugs be useful for patients with Parkinson's disease?

Q9 Are antimuscarinic drugs suitable for treating Parkinson's disease in the elderly?

Q10 When could the diagnosis of Parkinson's disease be confirmed?

Q11 What factors should the doctor consider in choosing an appropriate medication for this patient?

CASE STUDY 10 Rose's loss of consciousness

Learning outcomes

On completion of the following case study, you will be able to:

- understand the basis of epileptic seizures and the different categories, signs and symptoms of seizures;
- describe the pathophysiology of epilepsy and pharmacological approaches in managing the condition;
- gain knowledge of the mechanisms of action of the drugs available to treat epilepsy, particularly valproate, together with their side effects.

Part 1

Rose fell unconscious yesterday while taking a shower. She is 17 years old and this was the fourth time that she had lost consciousness in the past two weeks. She was hoping to start driving lessons but her condition has caused great concern and she postponed the lessons. Her mum took her to visit their family doctor. Following questioning, it was revealed that two of Rose's cousins also suffered from this condition and so did their grandfather, who died three years ago. The doctor referred Rose to the hospital for an electroencephalogram (EEG) recording. Rose was concerned about the process for recording the EEG, but the doctor explained the non-invasive nature of the method, and that put her mind at rest. The EEG recordings obtained from Rose's scalp showed abnormal electrical activity (spikes with sharp deflections and wave abnormalities). A diagnosis was made and valproate was prescribed.

Q1 What is the likely diagnosis of Rose's loss of consciousness?

Q2 What is epilepsy?

Q3 Comment on the underlying pathophysiology of epilepsy.

Q4 Comment on the uses of EEG.

Q5 What are the two main categories of epileptic seizures?

Q6 What are the signs and symptoms of seizures?

Q7 What type of seizure is consistent with Rose's symptoms?

Q8 Comment on the pharmacological management of epilepsy.

Q9 What is the mode of action of valproate and what side effects are associated with its use?

Q10 Give three examples of drugs recommended for treating epileptic seizures, commenting on their mechanisms of action.

Part 2

Rose was feeling better following the valproate therapy, which successfully controlled her symptoms. However, a few months later she realized that she might be pregnant and a pregnancy test later confirmed this. Rose had previously experienced very regular menstrual periods but that had changed in the last few months and was partly responsible for her unexpected and unplanned pregnancy. Rose decided to continue with the pregnancy but she was not sure whether to keep taking her medication or to stop, since she was worried about the health of the developing baby.

Q11 Could valproate affect the regularity of Rose's menstruation?

Q12 Could valproate harm the developing foetus?

Q13 Should Rose stop taking her medication during pregnancy?

Q14 What alternative medications are available for pregnant women with epilepsy?

CASE STUDY 11 Another day away from the office

Learning outcomes

On completion of the following case study, you will be able to:

- describe migraine in terms of its causes, types and associated symptoms;
- describe its pathophysiology and associated neurotransmitters;
- explain the treatments available to manage the symptoms of migraine.

Sue, who is 50 years old, works as an administrator in a busy office. She had to take the afternoon off again, in spite of the heavy workload in the office. She has been having severe headaches recently with visual disturbances; these make her feel nauseous, and sometimes vomiting occurs. She was treating the headaches with painkillers such as paracetamol/acetaminophen for a while; however, recently these drugs do not seem to help. Nobody in her family has suffered from any headaches like this, and she has never had any accident or injury to her head. She finally decides to see her family doctor. The doctor has now made a diagnosis and has prescribed sumatriptan.

Q1 What is the likely diagnosis of Sue's symptoms?

Q2 What are the usual symptoms of migraine?

Q3 Comment on the people at risk of this condition and its incidence.

Q4 Outline the causative factors that trigger migraine.

Q5 What are the two main types of migraine?

Q6 Describe the symptoms of the aura.

Q7 Comment on the underlying pathophysiology of migraine.

Q8 Which neurotransmitter is thought to play a role in mediating migraine?

Q9 Are all of Sue's symptoms consistent with the profile of migraine?

Q10 Comment on the treatments available for migraine.

Q11 To which category of drugs does sumatriptan belong?

CASE STUDY 12 Drooping eyelids

Learning outcomes

On completion of the following case study, you will be able to:

- describe the synapse and the classification of synapses;
- describe the processes involved in synaptic transmission at the neuromuscular junction;
- explain myasthenia gravis and its associated symptoms;
- describe causes which lead to the disease;
- outline pharmacological approaches to managing the symptoms.

Part 1

Mrs Downs has been to see her doctor complaining that she has been feeling weak lately and becomes tired very easily. She has also noticed that she has a problem focusing on objects and that her upper eyelids are drooping (this seems worse when she looks upwards). She feels that the muscles in her neck and shoulders are becoming very weak and she has difficulty in carrying heavy shopping. The doctor concludes that Mrs Downs has a neurological problem, makes her diagnosis and prescribes neostigmine plus atropine.

Q1 What is the likely diagnosis of Mrs Downs' symptoms?

Q2 What is myasthenia gravis?

Q3 Comment on the underlying pathophysiology of myasthenia gravis.

Q4 Are Mrs Downs' symptoms consistent with the profile of myasthenia gravis? Explain your answer.

Q5 What is a synapse?

Q6 How many types of synapse exist in the body?

Q7 Describe the events which occur at the synapse leading to the release of a neurotransmitter from the nerve terminal.

Q8 Do the number of synapses change with age?

Q9 To which category of drugs does neostigmine belong? Your answer should include its mechanism of action.

Q10 What is the rationale for using atropine in combination with neostigmine?

Part 2

After a few weeks, Mrs Downs went back to her doctor, complaining that her medication was not very effective.

Q11 Suggest another drug suitable for Mrs Downs.

3

Endocrine disorders

CASE STUDY 13 An agitated mother

Learning outcomes

On completion of the following two case study, you will be able to:

- describe the functions of the thyroid gland;
- explain the processes which control the secretion of thyroid hormones;
- list the symptoms of hyperthyroidism;
- outline the treatment of hyperthyroidism.

Mrs Kay has two active children and for the last year has become increasingly nervous and agitated, especially when dealing with them. She is constantly rushing from one task to another, always feels hot and seems to sweat a lot, even when the room temperature is quite low. She has a very good appetite but has lost some weight recently. In addition, she feels constantly tired and weak. She has been to see her doctor today. He thoroughly examined Mrs Kay, including her knee-jerk and other reflexes. He noted a swelling on her neck, rapid onset and relaxation of her reflexes and that her heart rate was higher than normal (87 beats per minute). During the consultation, it became clear that none of Mrs Kay's family had suffered any

Clinical Physiology and Pharmacology Farideh Javid and Janice McCurrie

problems of this kind. A blood test was taken and showed high levels of thyroxine (T_4) and triiodothyronine (T_3) of 30 $\mu g\,dl^{-1}$ and 300 $ng\,dl^{-1}$, respectively, compared with the normal values of 5–11.5 $\mu g\,dl^{-1}$ for T_4 and 100–215 $ng\,dl^{-1}$ for T3.

Q1 Where is the thyroid gland located in the body?

Q2 Which hormones are secreted by the thyroid gland? Which is the most active of the thyroid hormones?

Q3 Draw a flow chart showing the mechanisms that control thyroid hormone release from the thyroid gland.

Q4 Describe the characteristics of hyperthyroidism. Are the signs and symptoms experienced by Mrs Kay consistent with the profile of hyperthyroidism?

Q5 What is the most common cause of hyperthyroidism?

Q6 What is Graves' disease?

Q7 In addition to Graves' disease, what other conditions can cause hyperthyroidism?

Q8 Comment on the drugs used to treat hyperthyroidism. Your answer should include examples and their mechanisms of action.

Q9 Comment on the main side effects of carbimazole.

CASE STUDY 14 A vague and sleepy lady

Learning outcomes

On completion of this case study, you will be able to:

- describe the general effects of diminished thyroid function;
- list the symptoms of hypothyroidism;
- describe the condition of myxoedema;
- explain the significance of increased plasma thyroid stimulating hormone (TSH) concentration;
- outline the treatment of high plasma cholesterol.

Part 1

Zadie is a 70-year-old lady who has become increasingly vague and sleepy over the last year. She is depressed, finds it difficult to concentrate and her memory is failing. Often she wanders down to the local shop, forgets what she intended to buy and returns home empty-handed. Her fridge is often empty and she has given up cooking for herself, saying that she has no appetite. Zadie wears a thick jumper and a coat indoors. She has her heating on a very high setting as she always feels cold, even though visitors find it stiflingly hot in the house.

After much coaxing, her daughter persuades Zadie to visit the doctor as she suspects that her mother may be suffering from dementia. On examination, Zadie weighs 11 st (154 lb) an increase of 12 lb over her normal weight, although she has been eating very little. Zadie's thyroid gland is large and firm. Her face is puffy, her hands and feet feel cold to the touch and her reflexes are slow. The doctor suggests that Zadie may have a problem with her thyroid gland.

Q1 What is the relationship between the thyroid gland, the pituitary gland and the hypothalamus?

Q2 There may be several causes of fatigue and memory loss, for example dementia, severe anaemia and thyroid deficiency. What factors in the history suggest that this lady has a thyroid problem?

Q3 Make a summary of the actions of the thyroid hormones triiodothyronine (T_3) and thyroxine (T_4)

Q4 Why may Zadie have gained weight if she eats very little?

Q5 List the symptoms of hypothyroidism and outline the treatment of this condition.

Q6 What are the common causes of hypothyroidism?

Part 2

The results of Zadie's blood test are shown below.

	Test	Normal range
T_4 ($\mu g\,dl^{-1}$)	6.1	5–11.5
T_3 ($ng\,dl^{-1}$)	130	100–215
TSH ($\mu U\,ml^{-1}$)	20.5	0.7–7.0
Total plasma cholesterol ($mg\,dl^{-1}$)	246	140–220
Haematocrit (%)	40	38–44
Haemoglobin ($g\,dl^{-1}$)	13	12–16

TSH, thyroid stimulating hormone.

Zadie's blood pressure (BP) is 130/78 mmHg; heart rate 58 beats per minute.

Q7 Anaemia can cause some of the symptoms which Zadie reports, for example fatigue, sleepiness, lack of concentration. How does the blood test help to eliminate anaemia as a diagnosis?

Q8 Are Zadie's BP and heart rate normal?

Q9 What is the significance of the high TSH measured in this patient?

Q10 Many body processes, including thyroid function, are controlled by negative feedback mechanisms. Explain what is meant by the term *negative feedback*.

Q11 Does Zadie's blood cholesterol require treatment? If so, what drugs are available to reduce it?

CASE STUDY 15 A dehydrated businesswoman

Learning outcomes

On completion of the following case studies, you will be able to:

- describe the functions of the parathyroid glands;
- explain the processes which control its secretion;
- explain the pathophysiology and symptoms of hypo- and hyperparathyroidism;
- outline the treatment/pharmacological management of the condition.

Nazira has been a very successful businesswoman. She used to work long hours in her office but still managed to find time to go to the gym and swam four times a week. Recently, she has been feeling unwell and has become increasingly tired over the past three months. She decided to take a break from work for a couple of weeks. However, that didn't help much as she was feeling sick, dehydrated and very tired. She also developed polyuria. After examination and a blood test, her doctor noted that she showed an increased plasma level of calcium (hypercalcaemia), 5 $mmol\,l^{-1}$ as compared with the normal range of 2.2–2.55 $mmol\,l^{-1}$. Her blood glucose level was normal.

Q1 What is the likely diagnosis of Nazira's symptoms?

Q2 Which hormones are involved in the control of calcium balance in the body?

Q3 Where is the parathyroid gland located?

Q4 Comment on the function of the parathyroid gland and the actions of its hormone.

Q5 What is hyperparathyroidism and what are its consequences?

Q6 What causes hyperparathyroidism?

Q7 Which other conditions can cause hypercalcaemia?

Q8 Describe the major symptoms of hypercalcaemia.

Q9 Comment on the drug treatment of hyperparathyroidism.

Q10 Describe the condition of hypoparathyroidism.

Q11 Comment on the pathophysiology of hypoparathyroidism.

Q12 What are the effects of hypocalcaemia?

Q13 Outline the relationship between hypoparathyroidism and hypomagnesaemia.

Q14 Comment on the pharmacological management of hypoparathyroidism.

CASE STUDY 16 Brian's weight gain

Learning outcomes

On completion of the following case study, you will be able to:

- describe the location of the adrenal glands and the functions of the hormones secreted by the cortex;
- explain the processes which control the secretion of glucocorticoids and mineralocorticoids;
- explain the pathophysiology and symptoms associated with Cushing's syndrome and Addison's disease;
- outline the treatment/pharmacological management of these conditions.

Last year his computer business made record profits but this year Brian, a hard-working 45 year old, finds that both his life and his business have hit a low point. He has even failed to achieve basic sales targets. Brian is not getting on well with either employees or customers, feels miserable and thinks everything and everybody is against him. Brian has been ill several times recently with colds and other infections and feels constantly tired and weak. He has put on a lot of weight, mainly round his abdomen, neck and shoulders; his face now looks quite round and fat, but surprisingly his legs and arms seem to be getting thinner. Because he is aware of feeling more thirsty than usual (polydipsia) and needing to pass urine frequently (polyuria), Brian worries that he may be developing diabetes mellitus, and makes an appointment to see his doctor.

On examination Brian weighs 18 st (250 lb) and his BP is 160/95 mmHg. Taking Brian's appearance, mood change, BP and symptoms of polyuria and polydipsia into consideration, a provisional diagnosis of an adrenal gland disorder is made.

Q1 What is the structure of the adrenal glands and where are they located?

Q2 Which hormones are produced by the adrenal cortex? Your answer should include the functions of each hormone.

Q3 Which hormones are produced by the adrenal medulla and what is the function of the medulla?

Q4 How is the secretion of glucocorticoids regulated?

Q5 List the symptoms of Cushing's syndrome and Addison's disease.

Q6 Taking all the factors of the history into consideration, what is the likely diagnosis of Brian's illness?

Q7 Comment on the causes and treatment of Cushing's and Addison's disease.

Part 2

No abnormalities were found in Brian's urine and his blood cell count was normal. However, his blood glucose was 8.1 $mmol\,l^{-1}$ (normal value 3.5–6.7 $mmol\,l^{-1}$) and a glucose tolerance test later indicated impaired glucose tolerance. Tests for plasma insulin and thyroid hormones (T_3, T_4 and TSH) showed normal levels. Two further tests were then performed. A 24-hour urine sample was collected and Brian's free cortisol excretion was found to be considerably higher than normal. A second test, the dexamethasone suppression test, was also carried out. In this test, the patient is given a dose of dexamethasone at 11–12 p.m. and plasma cortisol is measured early next morning.

Q8 Brian showed impaired glucose tolerance and polyuria. In addition, some of his symptoms are similar to those observed in hypothyroid conditions. However, his problem is very unlikely to be due to either diabetes or hypothyroidism. Why is this?

Q9 Why is a 24-hour urine collection, rather than a single sample, necessary when estimating a patient's cortisol excretion?

Q10 What effects on cortisol secretion would you expect to observe following the dexamethasone test?

Q11 How is the secretion of aldosterone regulated, and what effects would you expect to observe in patients with increased secretion of this hormone?

Q12 How do you think Brian's medical condition may be related to his failing business and his changed attitude to employees and customers?

CASE STUDY 17 The thirsty schoolboy

Learning outcomes

On completion of the following two case studies, you will be able to:

- describe diabetes mellitus and the symptoms of this condition;
- explain the processes which control glucose in the body;
- explain the pathophysiology and pharmacological management of diabetes mellitus;
- explain the consequences of hypoglycaemia, hyperglycaemia and insulin overdose.

Ann was worried about her young son. Recently, he had not put on weight as expected in a growing child, and he appeared to be tired and thirsty all the time. Last week Ann received a report from his teacher, who noticed that he needed to go to the toilet more frequently than the other children in the class, and seemed to be tired and lethargic during school hours. He was often struggling to keep up during sports lessons in school. Ann had also noted that her son didn't go out to play for long; halfway through games of football with his friends he had to come home to rest. She became increasingly concerned and made an appointment with their family doctor. A simple urine test at the surgery showed an elevated level of glucose in his urine.

Q1 Which hormones are responsible for controlling the level of glucose in the body?

Q2 Where is the source of insulin and glucagon and what are the functions of these hormones?

Q3 What is the normal level of glucose in the blood during fasting?

Q4 Outline the major types of diabetes.

Q5 Describe the characteristics of type 1 diabetes. Your answer should include the group of people at risk, incidence of the condition and symptoms.

Q6 Explain how hypoglycaemia is associated with glycogenolysis and gluconeogenesis.

Q7 How are ketones involved in diabetes, and what are the consequences of ketoacidosis in the body?

Q8 Explain how lack of insulin can lead to atherosclerosis.

Q9 From which condition does Ann's son suffer?

Q10 Suggest another test that could be performed to confirm the diagnosis.

Q11 Comment on the pathophysiology of type 1 diabetes.

Q12 Comment on the pharmacological management of type 1 diabetes.

Q13 What precautions are necessary when taking insulin?

Q14 What is the treatment for insulin overdose?

Q15 How can we measure the overall success of a patient's blood sugar control?

CASE STUDY 18 Eric's expanding waistline

Learning outcomes

On completion of this case study, you will be able to:

- review the features of type 2 diabetes mellitus;
- outline common changes in the vasculature associated with diabetes;
- summarize renal, ocular and neuronal complications of poorly controlled diabetes;
- explain the uses of biguanide, sulfonylurea and thiazolidinedione drugs in the management of type 2 diabetes.

His workmates had started to tease Eric about his expanding waistline; the overalls he wore to work would no longer fasten round his middle. In fact, Eric was only too aware that his weight had increased by about 40 lb in the last three years to 200 lb, mostly as a result of daily takeaways and drinking copious amounts of beer each night in the local bar to 'cheer himself up' after his divorce. As he was only 5 ft 6 in. tall (1.65 m), not only did Eric look a mess, he felt awful too. He was now 55 years old, none of his clothes fitted properly, he was inactive but always tired and often thirsty, his sleep was interrupted each night by the need to urinate and he had recently suffered several episodes of skin infection and boils. So far he had managed to ignore his doctor's advice to try and lose weight, and today he was told that he needed to fast overnight and attend the clinic for a glucose tolerance test the following morning. How could he possibly cope without his nightly visit to the bar?

Q1 What is the likely diagnosis of Eric's symptoms? List the points in the case description which are significant for this diagnosis.

Q2 If Eric has the condition you describe, how will his glucose tolerance test differ from that of a healthy individual?

Q3 Describe type 2 diabetes, indicating how it differs from type 1.

Q4 What is the underlying pathophysiology of type 2 diabetes?

Q5 Changes to large blood vessels occur in poorly controlled diabetes. Outline these changes and comment on their effect on a patient's risk of cardiovascular disease.

Q6 Comment on the microvascular and renal changes which may occur in diabetes.

Q7 Describe the common nervous and ocular complications observed in poorly controlled diabetes.

Q8 Comment on the non-pharmacological management of type 2 diabetes.

Q9 Which categories of drug are used in treating type 2 diabetes? Your answer should include an example in each drug category you mention.

Q10 Describe the mechanisms of action and normal dosages of sulfonylurea and biguanide drugs.

Q11 What is the limiting factor in using sulfonylurea and biguanide drugs?

Q12 Explain how the mechanism of action of thiazolidinediones may help diabetic patients.

4
Cardiovascular disorders

CASE STUDY 19 Annie's heartache

Learning outcomes

On completion of this case study, you will be able to:

- describe the physiological characteristics of the coronary circulation;
- outline the autonomic control of the heart and the mechanisms which control coronary blood flow;
- review the causes of and possible treatments for angina;
- describe the mechanisms of action of the following anti-anginal drugs: nitrates, beta-adrenoceptor antagonists and calcium channel blockers.

Part 1

Annie is an elderly lady who lives with a one-eyed cat and a budgie. She is rather overweight and sometimes out of breath. Annie is very involved in the community, helps with many voluntary activities organized by the local Community Volunteers office and is generally considered quite active for a woman of over 70 years of age. Annie has had mild asthma for some years, but it troubles her very little and is well controlled.

Clinical Physiology and Pharmacology Farideh Javid and Janice McCurrie

However, Annie has noticed this winter that there is a tightness or ache in her chest when she walks back from shopping and climbs the hill to her house. On very cold days this is much worse and sometimes the sensation builds up from a dull ache to a real pain, which appears to spread down her left arm; it usually disappears after she has rested for a few minutes and had a cup of tea.

Annie's doctor diagnoses angina and prescribes glyceryl trinitrate (GTN) tablets, which are to be dissolved under the tongue, not swallowed.

Q1 What are the physiological characteristics of the coronary circulation?

Q2 Describe the innervation of the heart.

Q3 How is the blood flow to cardiac muscle controlled?

Q4 Explain why there is chest pain when Annie climbs the hill, especially in cold weather.

Q5 What electrocardiogram (ECG) changes would you expect to occur during an anginal episode?

Q6 GTN is commonly prescribed for angina. What is its pharmacological action?

Q7 What are the adverse effects of the nitrate drugs?

Part 2

Annie goes back to her doctor 18 months later. She reports that the pain in her chest is now coming more frequently. She has given up her voluntary work as she cannot now manage to walk up the hill and climb the stairs to the Community Volunteers' office. She also reports that she doesn't obtain adequate pain relief from the GTN prescribed. The doctor notes the worsening of her symptoms and suspects that she might not be obtaining pain relief from GTN because she is swallowing the tablets. He decides to prescribe her an alternative anti-anginal agent.

Q8 Explain the types and common causes of angina.

Q9 Why might swallowing the GTN tablets limit Annie's pain relief?

Q10 Would the beta-adrenoceptor (β-adrenoceptor) antagonist propranolol be suitable medication for Annie? Give reasons for your answer.

Q11 What is the pharmacological action of propranolol?

Q12 How can angina be distinguished from myocardial infarction?

Q13 A third type of agent available for the treatment of angina is a calcium channel blocker. Explain the pharmacological action of calcium channel blockers.

CASE STUDY 20 The executive's medical check-up

Learning outcomes

On completion of this case study, you will be able to:

- review mechanisms which control blood pressure (BP) and the ideal range for adult BP;
- explain the mechanism of action of agents currently available to treat hypertension;
- explain how antihypertensive agents may cause postural hypotension;
- outline lifestyle factors which affect BP.

Part 1

The chief executive of a European company, Sam Smart, is now 60 years old and has a high-profile, stressful job with frequent travel and business entertaining. He enjoys the good life and his weight has increased significantly over the years. Sam's wife tries to help him lose weight by preparing sensible meals, but he eats out so often that her efforts are without effect. However, Sam feels pretty fit on the whole, except for occasional mild asthma attacks, usually following a chest infection. This is well controlled with an inhaled beta-2-agonist (β_2 agonist).

Sam missed his check-up with the company doctor last year because of a delayed return from a business trip. At his most recent medical, two years ago, Sam's BP was a little higher than expected, 145/93 mmHg. At the time the doctor advised Sam to modify his lifestyle to help lower his BP. However, Sam's self-control was never very good and, although he tried to eat and drink sensibly for a while, he soon went back to his old ways.

Q1 What is the normal BP range for Sam's age group and what mechanisms maintain the BP in this range?

Q2 Does stress affect BP? How might Sam decrease his stress level?

Q3 If BP continues to rise and is not treated, what adverse effects (including tissue damage) may occur?

Q4 It is usual to start treating hypertension with either a beta-blocker (β-blocker) or a thiazide diuretic. What is the mechanism of action of the beta-adrenoceptor (β-adrenoceptor) antagonists (β-blockers)?

Part 2

Sam is available for his medical this year, but arrives late and in a rush. He has enjoyed a big lunch – steak and chips with deep-fried onion rings, followed by jam sponge pudding and custard. He has drunk most of a bottle of wine plus a brandy, and so is easily able to produce a urine sample. The company doctor checks Sam's weight and BP and sends blood and urine samples for analysis.

Sam now weighs 99 kg (220 lb), his urine appears normal but his plasma cholesterol is rather high at 6.5 mmol l^{-1} (ideal value <5.2 mmol l^{-1}). The first measurement of Sam's BP is 168/115 mmHg. After Sam sits quietly for 15 minutes, his BP decreases to 150/108 mmHg, a significant increase compared to the previous medical.

The doctor decides to prescribe Sam an angiotensin-converting enzyme (ACE) inhibitor, captopril. However, a few weeks later Sam is back, complaining about his throat and a dry, hacking cough, which disrupts his business discussions and, worse, annoys his wife!

The doctor is sympathetic: cough is sometimes a problem for patients taking captopril; she prescribes an alpha-adrenoceptor (α-adrenoceptor) antagonist, prazosin, instead. Sam has problems initially with orthostatic hypotension; however, after dosage reduction, prazosin is well tolerated and Sam's BP settles at 143/87 mmHg.

Q5 Explain why more than one measurement of a patient's BP is necessary, and outline factors which might have contributed to the difference observed between Sam's two BP readings.

Q6 Outline the major features of the renin–angiotensin system and its relationship to BP control.

Q7 Explain why Sam should be prescribed an antihypertensive agent; indicate factors (such as mechanism of action and Sam's history) which would be considered when choosing to prescribe an ACE inhibitor in preference to another type of antihypertensive agent.

Q8 Outline the locations at which alpha- and beta-receptors are found in the circulation and explain why alpha-blockers cause orthostatic hypotension.

Q9 If Sam had not been able to tolerate prazosin, or if this agent had not produced a satisfactory decrease in his BP, what other classes of antihypertensive agent are suitable for him?

Q10 Sam's final BP was 143/87 mmHg, is this an appropriate outcome of treatment?

Q11 Is Sam's lunchtime diet satisfactory for cardiovascular health? What lifestyle modifications are likely to be useful in lowering Sam's BP?

CASE STUDY 21 A hypertensive emergency

Learning outcomes

On completion of this case study, you will be able to:

- outline the risk factors for essential and secondary hypertension;
- explain how high BP may damage the body tissues;
- outline tests to clarify the causes of hypertension;
- compare the antihypertensive actions of beta-adrenoceptor (β-adrenoceptor) antagonists, agents acting on the renin–angiotensin system and short-acting agents such as labetalol;
- explain how very high BP may be treated.

Part 1

Billie is a 26-year-old computer officer in a busy organization. Her job involves hours of screen work and she also has an evening job at the offices of a supermarket chain. She originally attributed her bad headaches to the stress of doing two jobs. However, her headaches have become much more severe over the last five days and so, since she has also experienced visual disturbances, she decided to go to the accident and emergency department of her local hospital today.

On examination Billie's BP was 230/125 mmHg and she was sweaty and anxious with a heart rate of 120 beats per minute. This is extremely high for a young person and may indicate malignant hypertension or a hypertensive crisis. Billie described a family history of hypertension and admitted that her BP was 'quite high' a year ago, when she was prescribed a β-adrenoceptor antagonist (beta-blocker). Since it did not make her feel better, she stopped taking the beta-blocker (β-blocker) after a few months. It emerged that the drug had made her so tired that it was difficult to continue her evening job, and she needed the extra income. Billie smokes about 10 cigarettes a day and drinks moderately, consuming the equivalent of two bottles of wine a week. Her urine sample showed a moderate albumin content but no other abnormalities.

Q1 What is the normal range of BP and resting heart rate in Billie's age group?

Q2 List the risk factors for the development of essential hypertension. Is Billie likely to have this type of hypertension?

Q3 Is Billie's alcohol consumption and smoking likely to be contributing to her BP problem?

Q4 Which antihypertensive agents are suitable for treating young adults with hypertension?

Q5 What are the adverse effects of the β-adrenoceptor antagonists (β-blockers)? Is fatigue a common side effect of their use?

Q6 Although the cause of essential hypertension is unknown, there are several possible secondary causes which could lead to increased BP. Name three conditions which are known to cause secondary hypertension.

Q7 What investigations could be performed to clarify the cause of Billie's extremely high BP?

Part 2

Very high BP can affect the brain, causing visual disturbance, irritability, confusion and possibly epileptic seizures. Billie's high BP could be due to escalating essential hypertension, but in a young person it is likely to be secondary to another cause. Before an investigation begins, it is important to establish whether she is pregnant, taking prescribed medicines or is self-medicating. An examination of Billie's eyes showed some haemorrhage and exudate in her retina, indicating severe hypertension. In patients with dangerously high BP, pressure must be reduced gradually over several hours, with frequent measurements to confirm that pressure reduction is satisfactory. Once BP is in the 'safe' range, other tests can be performed to discover a possible explanation for the severe hypertension.

X-rays, angiography and scans confirmed that Billie's problem was renal artery stenosis, a condition most commonly seen in females of 20–50 years of age. Removal of the obstruction to renal blood flow is required to reduce BP permanently, but in the short term drug treatment will be needed to lower BP to an acceptable range.

Q8 Why is it necessary to find out if a young female patient is pregnant, taking prescribed medication or self-medicating?

Q9 Sodium nitroprusside can be used to rapidly reduce BP in hypertensive emergencies, but it is not suitable as a regular antihypertensive medication. Why is this?

Q10 Renal artery stenosis causes the kidney to become ischaemic. How does this result in high BP?

Q11 Which agents would be suitable to treat hypertensive patients who have a high renin level? Give reasons for your answer.

Q12 Labetalol is a short-acting antihypertensive agent which can be used in a hypertensive emergency. What is the mechanism of action of labetalol?

CASE STUDY 22 Harry Mann's bad day

Learning outcomes

Completion of this case study, will enable you to:

- revise the production of tissue fluid;
- review factors involved in the development of heart failure;
- describe how a heart attack may be differentiated from other causes of chest pain;
- review the signs and symptoms of congestive failure;
- review the mechanisms of action of diuretic agents.

Part 1

Harry Mann is the 64-year-old owner of a furniture removal business, Kwik Move. Kwik Move provides a good income, but Harry is ambitious. He is not a patient man: his quick temper makes his relationships with employees tricky. He has recently been stressed out by his van driver's frequent, unexplained absences. Harry works 12 hours a day, smokes continually, hasn't had a day off in 30 years and has no time for anyone with problems. On Monday, he arrives at work early after his usual greasy breakfast fry-up and a long wait in traffic. The van driver has not turned up again and Harry is furious. He decides to drive the van and move the client's furniture himself. Arriving at the client's house, Harry soon feels very unwell. He is weak and breathless, has a sharp pain in his chest and nausea; two antacid tablets do not help. He moves a couple of packing cases and his chest pain worsens. Fortunately, the house owner insists on taking Harry to the local hospital.

It may not be easy to tell immediately if a patient is suffering from angina, myocardial infarction or a relatively minor problem such as indigestion, but tests can aid diagnosis. The tests show that a heart attack is unlikely and there are few ECG abnormalities, but medical staff notice that Harry's ankles are swollen and his chest X-ray shows some cardiac enlargement. Harry's BP is found to be 170/98 mmHg. He is prescribed a thiazide diuretic and told to see his doctor the next day.

Q1 What tests could be performed to show if a myocardial infarction has occurred?

Q2 Are Harry's swollen ankles and cardiac enlargement significant?

Q3 How is tissue fluid formed and how does this process change in heart failure?

Q4 Is Harry's BP OK for his age, and is it related to his heart condition?

Q5 Will the thiazide drug affect Harry's BP?

Q6 What is the mechanism of action of the thiazides and how may they help patients with heart failure?

Part 2

Harry decides to slow down, working part-time at Kwik Move and eating more sensibly. He initially feels quite well on the thiazide, but after a time becomes increasingly breathless in daily activities. He is often nauseated and has lost his appetite, he suffers from abdominal discomfort and feels quite bloated. Harry's BP is now well controlled, but in view of his symptoms the doctor decides to start Harry on a loop diuretic (torasemide) rather than increasing the dosage of the thiazide.

Q7 How are Harry's symptoms related to heart failure?

Q8 Why is changing the diuretic more appropriate for Harry than increasing the dose of thiazide?

Q9 What is the mechanism of action of loop diuretics and why may they be more useful in congestive heart failure than the thiazides?

Q10 Do loop diuretics produce any adverse effects in the patient?

Q11 Should Harry receive any specific information or counselling on his medical condition and the new medication?

CASE STUDY 23 Grandpa's silence

Learning outcomes

On completion of this case study, you will be able to:

- outline the characteristics of the cerebral circulation;
- outline the symptoms and causes of stroke together with risk factors associated with this condition;
- explain changes in blood gas composition which may occur in cardiovascular disorders;
- outline the renal and respiratory compensatory mechanisms which return pH and blood gases to normal;
- describe some consequences of cerebrovascular disease.

Part 1

Sally and her grandpa were great friends: he sat smoking his pipe and listened to all her adventures each day after school. At teatime they often shared a large, double-cheese pizza or Sally brought in burgers and fries from the local takeaway.

Today Sally's grandpa didn't answer her excited greeting when she ran in to tell him about her trip to the zoo. She found him collapsed on the living-room floor, breathing but immobile. Fortunately, she knew what to do and the ambulance she called came quickly. Grandpa was rapidly transferred to intensive care at the local hospital. He had suffered a stroke and his blood gases and biochemistry were as follows:

pH	7.49
arterial PCO_2	32 mmHg (4 kPa)
arterial PO_2	99 mmHg
Hb saturation	98%

Q1 Stroke, or a cerebrovascular accident, is a serious disturbance of the brain's circulation (cerebral circulation) and is the third most frequent cause of death in Western countries such as the United Kingdom and United States of America. What are the two main causes of stroke?

Q2 Briefly, what are the physiological characteristics of the cerebral circulation?

Q3 Grandpa's pH is higher and his arterial PCO_2 lower than normal:

(1) What are the normal values for these two parameters?
(2) What type of acid-base disturbance is presented here?
(3) What is the likely reason for the low arterial PCO_2 and high pH presented in this case?

Q4 How may the kidneys compensate for this patient's acid–base disturbance?

Q5 Outline other possible causes of alkalosis.

Q6 How may the respiratory system compensate for alkalosis?

Part 2

Conditions such as severe atherosclerosis, thrombosis and rupture of a blood vessel or aneurysm can severely compromise the blood supply to brain tissue. Following a stroke, the processes which lead to irreparable damage to neurones in the brain take approximately 6–24 hours to complete. There is little time to confirm diagnosis and arrange scans or imaging to distinguish between brain haemorrhage and infarction. Some patients fortunately recover quite well, but in others the damage is extensive and irreversible. Grandpa's recovery was reasonably good, but he was left with some speech difficulties and weakness on the right side of his body.

Q7 Are Grandpa's pH and blood gases likely to return to normal as he is recovering from his stroke?

Q8 List the risk factors for stroke.

Q9 Several pharmacological agents could theoretically be used to promote cerebral blood flow and potentially improve the outcome from a stroke. Are thrombolytic agents or anticoagulants likely to be suitable treatments for all cases of stroke?

Q10 If a patient is thought to be at risk of a stroke, would an antiplatelet agent, such as aspirin, be suitable for stroke prevention?

Q11 Which part of Grandpa's brain is likely to have been affected by the stroke?

Q12 What lifestyle advice might it be appropriate to give to Grandpa during his recovery?

CASE STUDY 24 The gardener who collapsed on his lawn

Learning outcomes

On completion of this case study, you will be able to:

- describe myocardial infarction and some long-term effects on cardiac function;
- describe how myocardial infarction is confirmed;
- outline the use of ECG recording and defibrillation;
- outline the pharmacology of agents used following myocardial infarction;
- describe the development and treatment of pulmonary oedema.

Part 1

Charlie Green, a keen gardener, is particularly proud of his large and immaculate lawn. Although he is over 75 years old, has recently experienced some chest pain and has become increasingly short of breath in the last two years, he insisted on going out early one day to rake up fallen leaves from his lawn. Soon after going out, his wife found him collapsed and unconscious on the grass with a very feeble pulse. She immediately called an ambulance. Charlie's heart fibrillated on the way to hospital and he was resuscitated using a defibrillator. Charlie's ECG changes were found to be consistent with a diagnosis of myocardial infarction, and assessment of plasma creatine kinase-MB and troponin I and T later confirmed this diagnosis.

Q1 Define the term *infarct*.

Q2 Cardioversion, which involves a large direct current (DC) shock across the chest, is used when a patient's ventricles are fibrillating. Explain how this procedure aids patient survival.

Q3 Patients who suffer a cardiac arrest are usually given epinephrine (adrenaline) and external cardiac massage. If necessary, atropine is also administered. Explain the pharmacological actions of epinephrine and atropine on the heart.

Q4 Why were Charlie's plasma troponin and creatine kinase measured?

Q5 When there is a cardiac problem a 12-lead ECG is usually taken. Where are the electrodes placed and what do these leads measure?

Q6 What is the form of a typical ECG trace recorded from lead 2?

Part 2

Once the heart has stopped fibrillating, the patient is given oxygen via a mask and blood is taken for cell counts, glucose, lipids and cardiac enzymes. Diamorphine (for pain relief) and aspirin are given and, if there are no contraindications, thrombolysis can be started using either streptokinase or tissue plasminogen activator.

Q7 What is the purpose of administering the aspirin and what is its mechanism of action?

Q8 Explain why thrombolytic therapy may be necessary and the pharmacological actions of streptokinase.

Q9 What factors would contraindicate thrombolytic therapy?

Q10 Some of the patients who have a heart attack have previously suffered from angina pectoris. Define the term *angina pectoris* and describe ECG changes which are often associated with this condition.

Part 3

Several months after his heart attack, Charlie has a scheduled appointment with his cardiologist. He has been taking his aspirin regularly, together with a beta-blocker (β-blocker) for his angina. He reports that he is now very short of breath, particularly at night when he goes to bed. He can only sleep if he props himself up with several extra pillows. This suggests that Charlie has developed left heart failure: he is prescribed furosemide.

Q11 Explain why Charlie has particular difficulties when he lies down at night and why the extra pillows help him sleep.

Q12 Explain how furosemide can help Charlie's condition.

Q13 What lifestyle advice would you consider giving Charlie?

CASE STUDY 25 Hanna's palpitations

Learning outcomes

On completion of this case study, you will be able to:

- describe the location of the heart valves and origin of heart sounds;
- outline the pathway of excitation in the heart and correlate this with the normal ECG;
- describe atrial fibrillation, its relationship to rheumatic fever and the effect of abnormal mitral valve function on the circulation;
- describe the effects of sympathetic stimulation on the circulation;
- outline the pharmacology of digoxin and its use in the treatment of atrial fibrillation;
- explain the actions and uses of warfarin.

Part 1

Hannah is looking forward to her 70th birthday party with all her grandchildren, but as the day approaches she feels increasingly tired and unwell. She recently suffered from a chest infection, and today Hannah is very breathless (dyspnoea) and has palpitations (irregular heart beats); she decides to consult her doctor. From the medical records, Hannah's doctor knows that she suffered from rheumatic fever as a teenager and remembers noting a slight heart murmur when he previously examined her chest. He suspects that she has a problem with one of her heart valves and makes a provisional diagnosis of mitral stenosis.

Q1 Where is the mitral valve located?

Q2 Identify the other important valves in the heart and describe their location.

Q3 Summarize the pathway travelled by a red blood cell after it enters the inferior vena cava and moves through the heart to arrive finally in the aorta.

Q4 Where does the cardiac impulse, which starts each heart beat, arise? Describe the pathway followed by the cardiac impulse as it is conducted through the heart to excite the ventricles.

Q5 How does the ECG recorded from a standard limb lead correlate with the excitation processes in the heart?

Q6 What events in the cardiac cycle cause the first and second heart sounds?

Q7 What is the main cause of cardiac murmurs?

Part 2

Rheumatic fever is now less common in developed countries, but in Hannah's youth it was an important cause of heart disease. The illness is due to infection with beta-haemolytic (β-haemolytic) streptococci, which cause a sore throat. In some young people, the bacterium induces antibody-mediated autoimmune responses, which initiate inflammatory changes in joints and heart valves, particularly in the mitral valve. Inflammation thickens and may partly fuse the cusps of the mitral valve, making it narrow and unable to close properly in ventricular systole. Eventually, cardiac output decreases and the ventricle begins to fail from overwork. Since blood is not pumped effectively from the left ventricle, pulmonary congestion develops.

Hannah's ECG shows an abnormal heart beat: she has now developed atrial fibrillation. Hannah is prescribed digoxin and an anticoagulant, warfarin.

Q8 How may the normal ECG be changed in atrial fibrillation?

Q9 If the cardiac output decreases, the sympathetic nervous system is stimulated. What changes would you expect to observe in the circulation following stimulation of the sympathetic nervous system?

Q10 Explain how the defective mitral valve affects blood flow and pressure within the heart. Can these changes account for Hannah's dyspnoea?

Q11 What class of drug is digoxin and what is its pharmacological action in the circulation?

Q12 List possible adverse effects of digoxin.

Q13 Why does Hannah need an anticoagulant?

Q14 Explain the mechanism of action of warfarin. How rapidly does this drug take effect?

Q15 Should Hannah be given any special advice or monitoring now that she is taking warfarin?

5

Respiratory disorders

CASE STUDY 26 Moving to England

Learning outcomes

On completion of the following case study, you will be able to:

- describe the symptoms of hay fever;
- outline the stimuli that can cause release of histamine;
- describe the histamine receptor subtypes and their locations in the body;
- comment on the uses of antihistamines.

It has been nearly five months since 53-year-old Mrs Smythe moved from Florida to England at the end of February, following her husband's career change. Before the move she considered herself to be a very healthy middle-aged woman who enjoyed walking in the countryside. However, since coming to England she has been complaining about the persistent symptoms of a cold, with a runny nose and watery eyes for the last three months. Whilst she is still very excited about the move, she can now no longer go for her daily stroll since her symptoms are even more severe during walking. After visiting her family doctor she was prescribed fexofenadine (Allegra).

Clinical Physiology and Pharmacology Farideh Javid and Janice McCurrie

Q1 Is Mrs Smythe's problem likely to be a cold or can you suggest an alternative diagnosis?

Q2 What is hay fever? What are its associated symptoms?

Q3 To which category of drugs does fexofenadine belong?

Q4 Explain why fexofenadine was prescribed for this patient.

Q5 Which other stimuli can release histamine in the body?

Q6 List the histamine receptor types. Where are these receptors located?

Q7 What is the daily dose of fexofenadine and are there any potential side effects when using this agent?

Q8 By giving an example of an H_2-receptor antagonist, explain the pathophysiological conditions for which these drugs are used in the clinic.

CASE STUDY 27 The sneezing boy

Learning outcomes

On completion of the following case study, you will be able to:

- describe the pathophysiology of allergic rhinitis;
- outline the causes of allergic rhinitis;
- explain why antihistamines may be contraindicated in asthmatic patients;
- describe alternatives to antihistamines in treating allergic rhinitis.

Part 1

It's that time of year when 14-year-old Dean's symptoms trouble him most. He suffers from excessive sneezing, rhinorrhoea and nasal congestion. These symptoms make him very irritable, he cannot sleep properly, feels very fatigued and as a result is less focused on his school work. Since some important school exams are due to begin soon, his mother insists that Dean sees a doctor. His doctor prescribes azelastine hydrochloride. Before writing the prescription, the doctor checked Dean's medical notes and questioned him to make sure that he did not suffer from asthma.

Q1 What is the likely diagnosis of Dean's symptoms?

Q2 What is allergic rhinitis and what are the causative factors?

Q3 Comment on the pathophysiology of this condition.

Q4 What category of drugs can be used for perennial allergic rhinitis?

Q5 To which category of drugs does azelastine hydrochloride belong?

Q6 Why is it important that Dean's doctor checks whether he suffers from asthma?

Part 2

After a couple of weeks Dean returned to his doctor, complaining that his symptoms were persistent.

Q7 Is there an alternative medication for Dean's persistent symptoms?

CASE STUDY 28 Mandy's sleepover

Learning outcomes

On completion of this case study, you will be able to:

- describe major factors which affect breathing and review the pathology of asthma;
- explain the respiratory function tests which are useful in assessing the severity of asthma;
- review the pharmacology of agents used in treating asthma;
- appreciate the links between pathophysiology, pathology, pharmacology and the management of a common respiratory problem.

Part 1

Mandy is 13 years old and her asthma is usually quite well controlled with a 'reliever' and a 'preventer' medication. Her father and brother also have asthma. She enjoys staying with her friend Jane, who, unlike Mandy, has pet rabbits, rats and gerbils. Mandy was happy to stay for a sleepover party at Jane's house. But, waking at midnight wheezing and coughing, she realized that she had left her inhalers at home. Jane's mother heard her coughing and wheezing so took her to the accident and emergency department of the local hospital, by which time Mandy was distressed and very short of breath.

Q1 Which drugs are likely to be present in the 'reliever' and the 'preventer' inhaler?

Q2 What factors could account for wheezing, coughing and shortness of breath, which occur in asthma?

Q3 List the risk factors for asthma and the triggers which may exist in Jane's house.

Part 2

On arrival at hospital, although wheezing and breathless, Mandy could describe her usual medication to staff. A blood test showed that her arterial pH was 7.25 and lung function tests produced the results indicated below.

Mandy was taken to the ward and given nebulized salbutamol. After a while her wheezing and coughing diminished and she was able to go home next morning.

Q4 Mandy's forced vital capacity (FVC) and forced expiratory volume in one second (FEV_1) are shown below:

$$FVC = 2300\,ml$$
$$FEV_1 = 950\,ml$$

Why are these tests useful in asthma? What is the FEV_1/FVC ratio in this patient? Does this ratio indicate a restrictive or an obstructive condition?

Q5 Name an additional test which would be useful in assessing the severity of a patient's asthma.

Q6 What is the mechanism of action of beta-2-agonists (β_2 agonists), such as salbutamol, in the airways?

Q7 Outline the advantages of using a nebulizer rather than a breath-activated or dry-powder inhaler. Would you expect this medication to fully reverse Mandy's bronchoconstriction?

Q8 Why was Mandy's arterial pH lower than normal?

Q9 Is there any significance in the fact that Mandy could tell the staff about her asthma and her usual medication?

Q10 If salbutamol is not sufficiently effective, which other agents might be useful in treating an acute episode of asthma?

Q11 Spacer devices are often used for treating asthma when the patient is less than five years of age. Explain the function of a spacer.

Q12 Are there any therapeutic agents which might be particularly useful in the prophylaxis of asthma in children?

CASE STUDY 29 Bob and Bill's breathing problems

Learning outcomes

On completion of this case study, you will be able to:

- outline the major features of obstructive and restrictive pulmonary diseases;
- describe the use of selected respiratory function tests and interpret the data produced;
- discuss the usefulness of bronchodilator and steroid drugs in different forms of pulmonary disease;
- appreciate the effects of smoking on the respiratory system and the benefits of smoking cessation.

Bob and Bill are cousins of about the same age and both have lung problems. Bob has been in the building trade for nearly 40 years. He is still working. This work is heavy and dusty and mainly involves installing and removing insulation from buildings. In addition, he helps his son, who has a sand-blasting business, cleaning up the outside of old city buildings discoloured by traffic fumes.

Cousin Bill has already retired; he left his job at the steel works a few years ago when his health deteriorated. Bill has a fine life, watching television all day, going to the local pub most nights and smoking about 40 cigarettes a day. Bill also likes birds, he keeps a few racing pigeons in his loft and has a rather tatty green parrot which is often free to fly around the living room.

Part 1 Bob's problems

Bob recently made a huge effort to give up smoking on the advice of his doctor and has nearly succeeded. He has become increasingly short of breath over the last three years, which makes working difficult. He has developed dyspnoea and tachypnoea and his doctor has indicated that he is not really fit enough to continue working and should now retire.

Bob was referred to the local hospital for assessment of lung function. Results of his latest pulmonary function tests are shown below.

$FEV_1 = 2500$ ml

$FVC = 2700$ ml

Bob's blood test showed that his arterial PO_2 was 80 mmHg and the arterial PCO_2 was 34 mmHg.

Q1 What is the meaning of the terms *dyspnoea* and *tachypnoea*?

Q2 Explain the uses of respiratory function tests.

Q3 What is Bob's FEV_1/FVC ratio? And is it within the normal range?

Q4 What type of respiratory disease may produce the test results observed in this patient? Would a bronchodilator drug be useful for Bob?

Q5 Are Bob's blood gas tensions normal? If they appear abnormal, what explanation can you suggest for these values?

Q6 Could Bob's present lung condition be related to his occupation?

Part 2 Bill's problems

Bill has smoked for 45 years and has a chronic, productive cough. He is now very short of breath. He first noticed this when he was walking uphill, but now he is short of breath walking on the flat. In the last few years, he has had a couple of chest infections, which were successfully treated with antibiotics. His doctor prescribed an inhaler last year but it made little difference to his breathing. Bill's latest lung function test is shown below.
$FEV_1 = 1000$ ml
$FVC = 2700$ ml
Bill's blood test showed the following composition for his arterial blood:

> arterial PO_2 = 73 mmHg, arterial PCO_2 = 50 mmHg, pH 7.3, oxygen saturation = 84%.

Q7 What is the FVC/FEV_1 ratio for this patient? What type of lung disease may result in this ratio?

Q8 Describe the pathological changes which occur in the lungs of people with this type of disease.

Q9 Explain how the results of Bill's blood test differ from those expected in normal arterial blood and how these values might have been produced.

Q10 Are the parrot and the pigeons relevant to this case?

Q11 How would Bill's health benefit from giving up smoking cigarettes? What might help him to stop smoking?

Q12 Would it be useful to prescribe Bill a bronchodilator or a steroid inhaler?

CASE STUDY 30 A punctured chest

Learning outcomes

On completion of this case study, you will be able to:

- describe the physiological processes of inspiration and expiration;
- outline the anatomy and functions of the pleura;
- outline the effects of cigarette smoking on lung function;
- interpret data from selected lung function tests;
- describe the condition and effects of pneumothorax.

Part 1

Earlier today Brad tripped and fell on to a pile of wood he had removed from the old garden fence he was fixing. Brad is a keen home improver; however, at 63 years of age and smoking two packs of cigarettes or more each day, he is no longer young, fit or agile. Brad's chest was scratched and very bruised from his fall. He felt a bit breathless afterwards but was determined to finish the job. However, after a couple of hours his breathlessness became severe and the right side of his chest became very painful.

At the local hospital a careful examination of the right side of Brad's chest revealed a small puncture wound between the ribs, in addition to heavy bruising. A chest X-ray showed that his right lung had collapsed (atelectasis), although his left lung was still fully functional. A diagnosis of right pneumothorax was made, which was possibly due to penetration of the chest wall by a small nail from the fence. At this time Brad was dyspnoeic and cyanosed. Before administration of oxygen, his blood gases were as follows:

arterial PO_2 70 mmHg, arterial PCO_2 50 mmHg, pH 7.28.

In addition, his blood was found to contain 11% carbon monoxide.

Q1 Define the terms *dyspnoea*, *atelectasis*, *pneumothorax* and *cyanosis*.

Q2 How does air normally enter the lung during inspiration?

Q3 Describe the process of expiration.

Q4 Describe the location of the pleural membranes.

Q5 How does the entry of air via a puncture wound in the chest affect the lung?

Q6 Why did the puncture wound in Brad's chest have no effect on his left lung (which was shown to be ventilated normally)?

Q7 What factors could account for Brad's abnormal blood gas concentration and pH?

Part 2

Air was aspirated from the right side of Brad's chest and the lung re-inflated well. He was given antibiotics and advised not to smoke for a few days. He was also counselled to seriously consider giving up smoking altogether. A week later, at a check-up, respiratory function tests were performed and blood gases were measured again.

The results were as follows:

FEV_1 3.5 l
FVC 5.0 l
Peak flow 450 l min^{-1}
arterial PO_2 85 mmHg, arterial PCO_2 43 mmHg, pH 7.39.
Carboxyhaemoglobin was 9%.
(FEV_1, forced expiratory volume in one second; FVC, forced vital capacity)

Q8 Comparing Brad's respiratory function test results with those of a healthy person, are his test results now within the normal range for a 63-year-old male, 5 ft 7 in. (1.7 m) tall?

Q9 Is Brad likely to be showing the early stages of a restrictive or an obstructive lung condition?

Q10 Brad still has a low arterial PO_2. Is this likely to be related to his smoking habit?

Q11 Suggest a reason for the high level of carbon monoxide in Brad's blood.

Q12 Explain the effects of cigarette smoking on the lung.

CASE STUDY 31 Carmen's repeated respiratory infections

Learning outcomes

On completion of this case study, you will be able to:

- describe the innervation of sweat glands;
- describe a diagnostic test and comment on the treatment of cystic fibrosis;
- outline the pharmacology of bronchodilator drugs;
- describe the pulmonary and digestive problems of cystic fibrosis patients;
- review the actions of fat-soluble vitamins.

Carmen is a pale 20-year-old lady. Like many cystic fibrosis patients, she was diagnosed when she was a baby; her problem was suspected when she suffered repeated respiratory infections and failed to gain weight as expected. Patients with cystic fibrosis secrete a high concentration of NaCl in their sweat and this forms the basis of an early diagnostic test for the condition.

Both the lung and pancreas of cystic fibrosis patients are particularly affected by the condition. The major gastrointestinal problem is absent or scanty pancreatic enzyme secretion, which leads to intestinal malabsorption and steatorrhoea. Many of the patient's lung problems are due to secretion of abnormally sticky mucus, which blocks airways, leading to chronic lung disease. Accumulated mucus forms plugs which are difficult to clear from the lung and are susceptible to bacterial infection. Treatment of patients usually includes postural drainage (chest physical therapy), which involves percussion of the chest wall with cupped hands while the patient is in a 'head down' position, a demanding physical treatment. As the disease progresses, fibrous changes occur, there is destruction of airway walls and hypoxia develops.

Carmen recently developed a heavy cold and persistent cough with production of greenish sputum. She was now audibly wheezing and breathless. Carmen was taken into hospital for a few days for administration of intravenous antibiotics.

Q1 Describe the innervation of sweat glands.

Q2 How is the sweat test for diagnosis of cystic fibrosis performed?

Q3 Outline the pharmacological properties of the cholinergic agent pilocarpine and its action on sweat glands in the skin.

Q4 Describe the components of the cough reflex.

Q5 Would a cough suppressant be a useful addition to the antibiotics in Carmen's treatment?

Q6 Comment on the reasons for including physical therapy, such as postural drainage, in treatment of this condition.

Q7 List the classes of drug which may be used as bronchodilators. Is administration of a bronchodilator likely to be useful to Carmen?

Q8 While Carmen was in hospital, use of her normal inhaler was stopped and drugs were administered using a nebulizer. What is the advantage of using a nebulizer for this patient?

Q9 Are there any other therapeutic agents which could help Carmen's breathing?

Q10 Explain why patients with cystic fibrosis may fail to gain weight as expected in infancy.

Q11 Define the term *steatorrhoea* and give reasons for the presence of steatorrhoea in cystic fibrosis patients.

Q12 Carmen's recovery was good and on discharge from hospital she was prescribed vitamin supplements and a high-calorie, high-protein, extra-salt diet. Explain why this diet was considered suitable for Carmen.

Q13 Patients with cystic fibrosis do not readily absorb fat-soluble vitamins. Identify the fat-soluble vitamins and comment on the possible effects of poor absorption of these vitamins.

Q14 Cystic fibrosis patients are usually prescribed pancreatic enzymes to aid digestion. Which enzymes are likely to be included in these preparations?

CASE STUDY 32 Chandra's chronic bronchitis

Learning outcomes

On completion of this case study, you will be able to:

- define and describe the characteristics of *chronic bronchitis*;
- describe how lung function changes in bronchitis;
- differentiate between bronchitis and other *chronic obstructive pulmonary diseases* (COPDs);
- outline the causes and treatment of acute exacerbations of bronchitis;
- comment on the suitability of selected drugs in the treatment of chronic bronchitis.

Part 1

Chronic bronchitis, an obstructive pulmonary disease, usually results from repeated irritation or trauma to the lower respiratory tract, often as a result of smoking or prolonged working in dusty environments. Our patient, Chandra, is an elderly man with long-standing obstructive lung disease and a chronic cough, producing grey-coloured sputum. He spent many years as a miner before coming to the United Kingdom and now lives in a deprived area of the city in a damp, poorly heated flat.

Chandra's condition has progressively worsened and for the last three years he has suffered recurrent chest infections throughout the winter; these were treated with antibiotics. But his most recent chest infection did not respond well: he is wheezing, short of breath in daily activities, coughs frequently and is producing thick, green sputum. He was brought to hospital today because he is now very breathless at rest, is cyanosed and his lips look blue; his body temperature is 39 °C. At present he is too breathless to perform respiratory function tests, but a blood sample shows the following results:

arterial PO_2 72 mmHg, arterial PCO_2 55 mmHg, pH 7.3,
Hb saturation 81%, haematocrit 59%.

Q1 When Chandra's condition has improved, which respiratory function tests would you recommend and how do you expect the results to differ from normal values?

Q2 Describe the major types of COPD and compare the conditions you identify.

Q3 What factors in this history suggest that Chandra is suffering from an exacerbation of chronic bronchitis?

Q4 Patients with an obstructive disease often have to use expiratory muscles to help overcome expiratory airflow obstruction. Which accessory muscles are used and how does the shape and size of the chest eventually change in obstructive disease?

Q5 Define the term *haematocrit* and explain why it is elevated in this patient.

Q6 Explain why the patient now appears cyanosed. Is your explanation supported by the results of the blood test?

Part 2

Chronic bronchitis is associated with cigarette smoking and air pollution. The death rate from COPD increases with age, is higher in males than females, higher in towns than in the countryside and is greatest in people of low socio-economic status living in poor conditions. Persistent irritation of the airways by inhaled particulates results in hypertrophy of the mucous glands and increased mucus secretion. Mucus can accumulate and plug small airways. It is easily infected by bacteria, viruses and fungal spores. Some, but not all, bronchitic patients have an underlying bronchoconstrictor component in their condition and in many cases there is a chronic inflammatory component which also contributes to the airway obstruction.

Although Chandra was too ill to participate in respiratory function tests today; at a hospital admission last year, the following measurements were made:

Test	Result	Normal
FVC (l)	1.2	4.8–5.4
FEV_1 (l)	0.7	3.8–4.6
RV (l)	2.8	1.2
TLC (l)	6.4	5.8–6.0
T_{CO} (ml min^{-1} $mmHg^{-1}$)	15	25

FVC, forced vital capacity; FEV_1, forced expiratory volume in one second; RV, residual volume; TLC, total lung capacity; T_{co}, carbon monoxide transfer factor.

Q7 Why may people, who have never smoked, develop COPD?

Q8 Which of the results in Chandra's respiratory tests confirms that he has an obstructive condition?

Q9 Explain why Chandra's total lung capacity (TLC) and residual volume (RV) have increased above the normal value.

Q10 Why is a test of carbon monoxide transfer (T_{CO}) capacity useful?

Q11 Which organisms commonly cause exacerbation of chronic bronchitis in elderly patients?

Q12 Which antibiotics are likely to be suitable to treat Chandra's condition?

Q13 Chandra's blood gas composition is abnormal. One factor which contributes to this is ventilation–perfusion inequality. How is alveolar ventilation normally matched to perfusion in the lung?

Q14 After his recovery from this acute bronchitis, would this patient benefit from a trial of a bronchodilator or steroid inhaler?

Q15 Comment on the suitability of the following agents for Chandra at this time: (i) pure O_2, (ii) a muscarinic agonist drug, (iii) cromoglicate and (iv) low-dose aspirin.

6

Kidney and body fluid disorders

CASE STUDY 33 Greg's glomerulonephritis

Learning outcomes

On completion of this case study, you will be able to:

- describe the anatomy of the glomerulus;
- explain how tissue fluid is normally formed and factors which may cause oedema;
- outline the mechanisms which normally control body fluid volume;
- describe common renal function tests.

Part 1

Greg is nine years old and is usually an active, cheerful boy who loves to play football. Over the last week his mother noticed that Greg was not as lively as usual and had no interest in going out to play. The only medical problem affecting Greg recently was

Clinical Physiology and Pharmacology Farideh Javid and Janice McCurrie

a nasty sore throat, which cleared up two weeks ago. But as he now looked pale and his face was rather puffy, his mother took Greg to the doctor. On examination, the doctor noted some abnormal chest sounds, suggesting that fluid had accumulated in Greg's lungs. Specimens of Greg's urine were taken for analysis. The small volume of urine produced appeared dark pink and frothy, tests showed the presence of red blood cells (haematuria) and a large amount of protein (proteinuria).

Q1 Describe the structure of the glomerulus and Bowman's capsule.

Q2 Which substances are normally filtered and pass into the proximal tubule? Which substances are usually unable to enter the proximal tubule?

Q3 How is tissue fluid normally formed in the body?

Q4 Greg had a significant amount of protein in his urine. How may this affect tissue fluid formation? Can it account for the puffy look and the fluid accumulation in Greg's lungs?

Q5 What factors other than kidney problems may cause oedema?

Part 2

Greg's illness was diagnosed as post-streptococcal glomerulonephritis. In some children, one to three weeks following a throat (or sometimes skin) infection with a beta-haemolytic (β-haemolytic) *Streptococcus*, the antibodies produced in response to the bacteria combine with bacterial antigens to form complexes that become trapped in the glomerular capillaries. Patients may become oedematous, drowsy, dyspnoeic and constantly thirsty. Kidney function is greatly diminished for a period, but most children eventually make a good recovery.

Q6 What is the definition of *dyspnoea* and how may it have developed in this patient?

Q7 How is the volume of extracellular fluid normally regulated?

Q8 How may the formation of antigen–antibody complexes cause a problem in the glomerulus?

Q9 Why do you think that Greg was constantly thirsty?

Q10 What treatments may be useful for Greg's condition?

Q11 Which tests could be applied to estimate whether Greg's kidney function had returned to normal?

CASE STUDY 34 Kevin's chronic kidney problems

Learning outcomes

On completion of this case study, you will be able to:

- describe the physiology of glomerular filtration;
- outline the basis of renal clearance tests;
- explain the significance of abnormal concentrations of albumin, urea, electrolytes and other substances in blood;
- describe how abnormal electrolyte concentrations may affect nerve, skeletal and cardiac muscle;
- describe how the normal and failing kidney affects the red blood cell count.

Part 1

Kevin is a mature student who has been an insulin-dependant diabetic since childhood. He has regular check-ups for his diabetes at the local health centre. Today at his appointment he admits that he has not felt well for the past two months. He complains of extreme tiredness, weakness, nausea and sometimes vomiting. He originally thought this was due to the stress of taking his final university examinations, which are about to start. But now he feels that maybe some other factor is responsible for these symptoms.

The practice nurse is concerned by Kevin's symptoms, especially since she notices that Kevin's ankles are swollen. She suspects that Kevin has developed a kidney problem. If this is quickly detected and treated, further damage and loss of kidney function may be prevented or at least slowed. Care in the control of blood glucose and blood pressure is very important for diabetic patients and can reduce kidney problems, such as loss of albumin in the urine. Urine and blood samples are taken, with the following results:

	Test sample	Normal range
Sodium (mmol l^{-1})	139	135–145
Potassium (mmol l^{-1})	5.3	3.5–5.0
Urea (mmol l^{-1})	45	3–6.5
Creatinine (μmol l^{-1})	580	50–120

Q1 What are the physiological characteristics of glomerular filtration?

Q2 How are the large amount of albumin in the urine and Kevin's swollen ankles related to the damaged glomerular membrane?

Q3 Is the very high creatinine concentration in Kevin's blood significant and which test of kidney function uses measurements of creatinine in blood and urine?

Q4 How is the potassium concentration of the body fluids controlled?

Q5 Can the increased concentration of urea in Kevin's blood account for any of his present symptoms?

Part 2

From the symptoms and examination of blood and urine, a diagnosis of chronic renal failure is made. Unfortunately, considerable kidney damage can occur, often over a period of years, before the patient notices the symptoms associated with chronic renal failure. As the amount of functioning kidney tissue decreases, blood electrolytes begin to change. At the same time, the ability of the kidney to excrete nitrogenous waste decreases and urea concentration in the blood rises (uraemia). The patient may remain symptom-free until the concentration of urea rises sufficiently to cause the nausea and vomiting Kevin has recently experienced.

Further results from Kevin's blood test are shown below:

	Test sample	Normal range
Red cell count ($\times 10^{-9}\,l^{-1}$)	3.9	4.5–6.5
Haemoglobin ($g\,dl^{-1}$)	7.6	13.5–17.5
Calcium ($mmol\,l^{-1}$)	1.6	2.2–2.55

Q6 From the test results given in Part A, Kevin appears to be suffering from hyperkalaemia. What effects can hyperkalaemia have on nerve and muscle tissue. Does the high $[K^+]$ in Kevin's blood account for any of his symptoms?

Q7 How does the kidney affect blood calcium concentration?

Q8 How do thiazide and loop diuretics affect the excretion of calcium?

Q9 Both red cell number and haemoglobin concentration are reduced compared to normal values. What are the signs and symptoms of anaemia?

Q10 Where are red blood cells produced in the adult and how is their production usually controlled? How can kidney failure affect the red cell count?

CASE STUDY 35 The polar bear's fun run

Learning outcomes

On completion of this case study, you will be able to:

- describe the control of body water and sodium balance;
- discuss the characteristics of the sensation of thirst;
- describe the actions and regulation of antidiuretic hormone (ADH), aldosterone and atrial natriuretic peptides;
- describe the responses of the body to dehydration and the effects of rapid water consumption on extracellular and intracellular fluids.

Part 1

Eddie was persuaded to join a sponsored fun run to collect for a children's charity. The last time Eddie had exercised seriously was many years ago at college, but he had no time for training sessions. Promises of sponsorship money were difficult to obtain and so, to appear more exciting to sponsors, Eddie decided to run in a costume, choosing a polar bear suit. On the day of the run Eddie discovered that his furry outfit was much hotter than expected and the day was unusually sunny. He soon became very uncomfortable and fell behind the other runners. At the halfway point he was extremely hot, sweaty and thirsty and quickly drank several bottles of water. After running a bit further he became disorientated and dizzy, wandered off the course and collapsed. He was discovered by the organizers, who removed the suit to cool him and gave him a long drink of water. But Eddie's temperature remained high, his head ached, his pulse raced and he still seemed dizzy and confused, so he was taken to the local hospital.

Q1 What factors promote the sensation of thirst and how is thirst related to the regulation of body water content?

Q2 Eddie lost a large amount of salt and water because of excessive sweating in his furry costume. How is sodium intake and sodium loss in sweat and urine related to body water balance?

Q3 How is the sodium content of the body normally regulated?

Q4 What is the effect on blood volume and pressure of a significant decrease in both salt and water in the body?

Q5 What could account for Eddie's high pulse rate and dizziness?

Q6 What changes in the osmotic pressure of intracellular and extracellular fluids would you expect to occur following ingestion of a large volume of water?

Part 2

Staff at the hospital reduced Eddie's temperature and, not realizing how much water he had recently taken, set up an intravenous (IV) drip of 5% dextrose in order to rehydrate him. He was not producing urine, and gradually became weaker, irritable and more confused. His skin looked puffy and oedema in dependent parts of his body soon developed. A blood sample showed the following:

	Test sample	Normal range
Sodium (mEq l^{-1})	125	135–145
Haematocrit (%)	<40	47 ± 7 (in male subjects)

The infusion of dextrose was stopped and fortunately after a time Eddie's kidney function improved; he produced a good flow of urine and very gradually recovered.

Q7 What is the significance of the low sodium concentration and haematocrit in Eddie's blood sample?

Q8 How does the kidney respond to dehydration following fluid loss from the body?

Q9 Where is ADH produced, how is it controlled and what is its action on the kidney?

Q10 Some patients produce an excessive amount of ADH, and in some cases production is reduced or the kidney fails to respond to the ADH produced. What are the likely effects of these conditions on urine production?

CASE STUDY 36 The housewife who drank too much

Learning outcomes

On completion of this case study, you will be able to:

- outline the sources of water gain and routes of water loss by the body;
- review the secretion and actions of vasopressin (ADH);
- describe mechanisms in the kidney which allow production of a concentrated or diluted urine;
- describe the features of diabetes insipidus.

Part 1

Irene always seems to be very thirsty. She is a 30-year-old housewife whose symptom has troubled her for some years. She now reports chronic, intense thirst and excessive drinking (polydipsia) with polyuria. Irene is losing weight, she has dry mucous membranes and skin even though she drinks large quantities of water and juice all day. Her excessive fluid intake and a constant need to urinate, even at night (nocturia), makes her constantly tired and irritable. Her family is worried and persuade her to visit the doctor.

A sample of Irene's urine was tested for glucose: this test proved negative. A blood sample was sent off for determination of plasma vasopressin (ADH), which appeared to be within normal limits. Irene was also asked to collect a 24-hour urine sample for analysis. Her 24-hour urine volume was 6.5 l.

Following a water-deprivation test, Irene's urine was found to have an osmotic pressure of 460 mOsm kg^{-1} water. The normal kidney can concentrate urine to approximately 1000 mOsm kg^{-1} water. A diagnosis of diabetes insipidus was made.

Q1 Define the terms *polydipsia* and *polyuria.*

Q2 List the routes and volumes of water gained and lost by the body. Approximately, what volume of urine is normally produced per day?

Q3 Describe the secretion of vasopressin and the stimuli which cause its release.

Q4 How is a concentrated urine produced and which parts of the nephron are sensitive to the actions of vasopressin?

Q5 Why is Irene showing signs of dehydration when she is drinking large quantities of fluid each day?

Part 2

There are two types of disturbances in vasopressin secretion. In central diabetes insipidus, vasopressin secretion is reduced: it can be treated by giving vasopressin or desmopressin, which has a longer half-life, by mouth or intranasally. In nephrogenic diabetes insipidus, the plasma vasopressin concentration may be normal but the kidney fails to respond. The latter type of diabetes insipidus does not respond to vasopressin therapy but, paradoxically, can be managed by giving a thiazide diuretic, for example chlortalidone, at a maintenance dose of 50 mg daily.

Q6 From which type of diabetes insipidus is Irene suffering? Give reasons for your answer.

Q7 If a patient is producing a large amount of vasopressin from a tumour, from what symptoms are they likely to suffer?

Q8 What conditions could result in disorders of vasopressin secretion?

Q9 Irene was treated with chlortalidone, a thiazide diuretic. For what other conditions are the thiazides usually prescribed?

Q10 Why are thirst and frequent urination prominent symptoms of both diabetes insipidus and diabetes mellitus?

7

Blood disorders

CASE STUDY 37 An exhausted mother

Learning outcomes

On completion of this case study, you will be able to:

- review the recycling of breakdown products of red blood cells;
- review the types and functions of white blood cells;
- list the causes, classification, signs and symptoms of anaemia and explain their physiological basis;
- review the changes in blood cells which distinguish the different anaemias;
- describe how the different anaemias are usually treated.

Part 1

Maria is in her late thirties, she has five healthy school-age sons. Her husband is an airline steward who is regularly away from home, so Maria copes with both housework and children by herself. Over the last few years Maria has felt tired much of the time. After moving house recently, the situation has become much worse; she

Clinical Physiology and Pharmacology Farideh Javid and Janice McCurrie

feels very weak and moody and absolutely dragged down by tiredness. Life is getting her down and nothing seems to help. She visits her family doctor, asking for a tonic.

Maria's medical notes show that she has mild rheumatoid arthritis (an autoimmune disease), the condition is not severe and her joint pain is well controlled by a non-steroidal anti-inflammatory drug (NSAID), taken twice a day. She is not trying to diet and apparently has a well-balanced food intake, but reports that she has recently had several minor stomach upsets and colds. A blood sample is taken for analysis, including a differential white blood cell count.

Maria's blood test results are as follows:

	Test	Normal value
Haematocrit (%)	34	38–45
Haemoglobin ($g\,dl^{-1}$)	10	12–16
Red cell count (million mm^{-3})	3.6	4.2–6.0
Mean corpuscular volume (μm^3)	105	81–96
Erythrocyte sedimentation rate ($mm\,h^{-1}$)	30	5–10
White blood cell count (mm^{-3})	10 000	4000–10 000

Q1 What would be the likely diagnosis of Maria's symptoms according to the blood analysis results?

Q2 What materials are needed to produce red blood cells with a haemoglobin content within the normal range?

Q3 What happens to the breakdown products of red cells when they are destroyed via the reticuloendothelial system?

Q4 Describe the different types of white blood cell and their functions.

Q5 What are the symptoms of anaemia and how are anaemias usually classified?

Q6 What is the diagnostic value of a differential white blood cell count?

Part 2

Anaemia occurs when there is a decrease in haemoglobin below the appropriate level for the age and sex of the individual. The anaemia may be due to several factors as lack of iron, vitamin B_{12} and folic acid all affect red cell production, resulting in anaemia. B_{12} deficiency may also cause neurological problems, such as numbness and weakness. Patients with B_{12} deficiency may also report mood swings and seem to suffer more infection and mild gastrointestinal problems than normal, so Maria's moodiness, stomach upsets and colds may be significant.

Q7 Maria's white cell count is quite high. Does this correlate with anything in her medical history?

Q8 Is Maria's anaemia likely to be caused by a dietary deficiency? How may anaemia be caused by a deficiency of folate and B_{12}?

Q9 Are folate and B_{12} equally effective forms of therapy for anaemia?

Q10 Using the blood test results and history, what factors were important for identifying the type of anaemia that Maria is suffering from in this case?

Q11 What is the diagnostic significance of the erythrocyte sedimentation rate (ESR)? Is this test useful in diagnosing anaemia?

CASE STUDY 38 Patsy's Australian journey

Learning outcomes

On completion of this case study, you will be able to:

- describe the process of intravascular coagulation;
- list the risks and symptoms of pulmonary embolism;
- outline the features of the fibrinolytic system;
- outline the pharmacology of thrombolytic agents.

Part 1

Patsy is a young mother flying home to London after visiting family in Australia. The trip was planned two years ago but was postponed when Patsy became pregnant. Now her baby is nine months old she has decided to leave him with her parents, while she and her partner make the trip. The flight is long and the conditions cramped in economy class. Patsy is particularly uncomfortable because the elderly passenger by her side is fast asleep, making it difficult for her to move without waking him. Although she is hot, thirsty and uncomfortable, Patsy thinks it unkind to disturb her sleeping neighbour.

Patsy is relieved when the plane lands so she can have a drink and a much-needed cigarette. However, while collecting their suitcases in the airport, she feels a sudden sharp pain in her chest. The pain is intense and does not diminish when she rests. She finds it difficult to breathe, her heart is racing and she collapses. Her partner is frantic at this sudden collapse with intense chest pain – is Patsy having a heart attack? Airport staff are very efficient and quickly take her to the airport doctor, who recognizes her problem. He diagnoses a pulmonary embolism and administers oxygen before organizing hospital admission.

Q1 What are the links between long-haul flights and the formation of emboli in the circulation?

Q2 What other factors in this history might predispose Patsy to intravascular clotting?

Q3 In which part of the circulation do the emboli, which finally lodge in the lung, usually form? Briefly explain why the area you have indicated is vulnerable to intravascular clotting.

Q4 How is vitamin K involved in the mechanism of blood coagulation?

Q5 Briefly explain the major steps in the intravascular coagulation process.

Q6 What is the difference between a thrombus and an embolus?

Part 2

Embolism associated with long flights is generally due to thrombus formation in deep leg veins (deep-vein thrombosis, or DVT). The thrombus may move to the pulmonary circulation, where effects on lung function depend on the extent of the blockage produced. A massive embolus may occlude the main pulmonary artery, resulting in hypotension, shock and possibly death; multiple small emboli cause little problem and are lysed by the fibrinolytic system. Sometimes surgical removal of the embolus is necessary, but in Patsy's case clot lysis was successful and she made an uneventful recovery.

Q7 What happens to the area of lung whose circulation has been cut off by an embolus in a pulmonary blood vessel?

Q8 Describe the system which both prevents coagulation and breaks down thrombi in the circulation.

Q9 Which drugs might be suitable to lyse emboli that have already formed in a patient's pulmonary circulation?

Q10 Venous stasis and embolism can be a problem in hospitalized patients. Which drugs may be useful to reduce the risks of embolization in such patients?

Q11 What precautions would be suitable for travellers to take when embarking on long-haul flights in order to minimize their risk of DVT and pulmonary embolism?

CASE STUDY 39 The dizzy blonde

Learning outcomes

On completion of this case study, you will be able to:

- outline the characteristics and causes of iron-deficiency anaemia;
- list the changes in skin, hair, blood and red cells which occur in iron-deficiency anaemia;
- outline the production of red blood cells and describe how iron is absorbed in the intestine;
- describe the treatment of iron-deficiency anaemia;
- describe the effects and treatment of iron poisoning.

Since she had gained weight after each pregnancy, Lizzie decided to start a very low-calorie diet after her youngest daughter was born. Instead of making her feel better, her weight loss has been accompanied by increasing weakness and fatigue. Her other three children, all under five years of age, are very lively and Lizzie feels permanently exhausted, often very dizzy and sometimes breathless; she occasionally suffers from palpitations when climbing the stairs. Lizzie's appearance has deteriorated lately: she is very pale, has sore patches round her mouth and her once-glossy blonde hair is dry and brittle and has started falling out. At the health centre it was noted that Lizzie had pale skin and conjunctivae, red areas at the corners of her mouth and a sore tongue, which, Lizzie explained, makes it uncomfortable to chew food, particularly meat. Her heart rate was 95 beats per minute and blood pressure (BP) was 95/60 mmHg.

Her blood test shows the following:

	Test	Normal range
Haemoglobin ($g\,dl^{-1}$)	8	12–16
Haematocrit (%)	30	38–45
Mean corpuscular volume (μm)	60	80–96
Reticulocytes (%)	6	0.8–2
Ferritin ($\mu g\,l^{-1}$)	5	15–200

Q1 Define the term *anaemia* and identify the type of anaemia from which Lizzie appears to suffer.

Q2 List the factors in the history that could suggest or contribute to this type of anaemia.

Q3 Which results in the blood test support a diagnosis of iron-deficiency anaemia?

Q4 What changes occur in blood as anaemia develops?

Q5 How is iron absorbed by the body and what factors can affect this process?

Q6 Where are red blood cells formed in the adult, how is their production controlled and how long do they survive in the circulation?

Q7 What are the causes of iron-deficiency anaemia and how does the deficiency of iron affect formation of red cells?

Q8 Should any other tests be made before Lizzie's treatment is started?

Q9 What is the usual treatment for iron-deficiency anaemia?

Q10 Are any side effects associated with this treatment? If so, are alternative forms of therapy available?

Q11 Iron preparations are a common cause of accidental poisoning in very young children, who may mistake iron tablets for sweets. How is acute iron toxicity treated?

8 Gastrointestinal disorders

CASE STUDY 40 Mr Benjamin's bowel problem

Learning outcomes

On completion of this case study, you will be able to:

- outline the defecation reflex;
- describe the factors that contribute to the development of constipation;
- discuss the pharmacological management of constipation;
- outline the side effects associated with the use of laxatives.

Mr Benjamin is a 75-year-old man with no close relatives who has lived alone since he lost his wife three years ago. He has become increasingly frail over the past two years. He cooks infrequently, eats little fruit and almost no vegetables. Mr Benjamin rarely visits his friends or the shops; if he goes for a walk, it is a short one, as he is now frightened of the traffic. He has severely restricted his intake of fluids in the evening and has even cut out his cup of hot milk before bed, as he does not want to visit the toilet during the night. Mr Benjamin has never had any gastrointestinal complaints in the past, but recently he has not opened his bowels for more than two weeks. His doctor has advised him to drink more fluids and has prescribed lactulose.

Clinical Physiology and Pharmacology Farideh Javid and Janice McCurrie

Q1 Describe the normal process of defecation.

Q2 What are the causative factors for the development of constipation?

Q3 Outline the factors in the history which may be contributing to this patient's constipation.

Q4 Which types of drug may cause constipation as a side effect?

Q5 Comment on the pharmacological management of constipation.

Q6 To which category of drugs does lactulose belong? Comment on its mechanism of action and the recommended dose.

Q7 Comment on the adverse effects which are associated with the use of laxatives.

Q8 What advice might be useful for this patient?

CASE STUDY 41 A disturbed holiday

Learning outcomes

On completion of this case study, you will be able to:

- define the term *diarrhoea*;
- describe the causes of diarrhoea and the organisms frequently associated with traveller's diarrhoea;
- discuss the drug treatments used in diarrhoea and their mechanisms of action;
- describe the actions of intestinal flora.

Sixty-year-old Mrs Kaye was a very healthy lady who never missed her daily walk to the park and went swimming twice a week. She had never experienced any health problems, except occasional indigestion, for which she usually took ranitidine. Sadly, a year ago she lost her husband, who died quite suddenly. Mrs Kaye sold her house and moved in with her daughter and two grandchildren. She decided to take them all abroad for the first time, to enjoy a package holiday in the sun. Everybody was having an enjoyable time; however, two days before coming back home, Mrs Kaye developed acute diarrhoea. Her daughter took her to a local medical centre and Mrs Kaye was prescribed loperamide hydrochloride. She had a rather uncomfortable few days but recovered soon after returning home.

Q1 What do you understand by the term *diarrhoea*?

Q2 Comment on the pathophysiology of diarrhoea.

Q3 Which organisms are most frequently associated with traveller's diarrhoea?

Q4 What factors might have made Mrs Kaye more likely than some other travellers to develop diarrhoea?

Q5 Comment on the pharmacological management of diarrhoea.

Q6 To which category of drugs does loperamide hydrochloride belong and what is its mechanism of action?

Q7 What is an intestinal flora modifier? Comment on its mechanism of action.

CASE STUDY 42 Jude's sudden admission to hospital

Learning outcomes

On completion of this case study, you will be able to:

- review the exocrine and endocrine secretions of the pancreas;
- outline the diagnostic tests for pancreatitis;
- describe the control of pancreatic secretion;
- explain the effects of acute and chronic pancreatitis.

Part 1

Jude is a 22-year-old woman who was taken to hospital one morning complaining of knife-like, severe upper-abdominal pain, nausea, vomiting and fever. She stated that she had been feeling quite well until this morning, when her symptoms came on suddenly. On questioning, the patient admitted that she routinely consumed very large amounts of alcohol and that she had drunk rather more than usual during the preceding two weeks as she was 'celebrating'. In addition, she also revealed that her boyfriend had recently been admitted to hospital suffering from hepatitis.

Q1 What is likely to be the initial diagnosis of Jude's symptoms and why?

Part 2

On further testing, a specimen of urine showed a normal colour and the patient did not appear to be jaundiced. In addition, a CT (computerized tomography) scan showed that her liver and bile duct were normal, and she had no history of stomach (peptic) ulcers.

Q2 Are the results in Part 2 consistent with the initial diagnosis? Give reasons for your answer.

Q3 What signs could indicate that Jude was jaundiced and why is the colour of her urine significant?

Part 3

On admission to the ward Jude had blood samples taken for biochemical tests. Her serum amylase activity was markedly increased and after a day her lipase was also significantly raised. The pattern of change in Jude's enzymes was as follows:
amylase rose quite quickly over the 3–12 hours after onset of symptoms and fell back to the normal value in three to four days; this rise was sharp and prominent. The increase in lipase was slower and more prolonged, lasting about seven days. Jude was given pain relief, and an intravenous catheter was inserted for a glucose and saline drip, with orders that she be given nothing by mouth.

With appropriate pain control, Jude soon felt better and her recovery was complete.

Q4 What is the ultimate diagnosis of Jude's symptoms?

Q5 Why do you think there was an instruction to give Jude nothing by mouth? And why was a mixture of glucose and saline given intravenously?

Q6 What is the function of amylase in the human gut and which other gut structures produce an amylase?

Q7 List the digestive enzymes (at least five) which are produced by the pancreas.

Q8 In pancreatitis the pancreas is damaged by some of these enzymes. Why is the normal pancreas *not* affected by the enzymes it produces?

Q9 Outline the actions of amylase on the gut contents, naming the products of digestion.

Q10 What factors normally control pancreatic enzyme secretion?

Q11 Comment on the tests which aid diagnosis in suspected pancreatitis.

Q12 The pancreas has both endocrine and exocrine cells. What is the difference between endocrine and exocrine secretions?

Q13 Which hormones are produced by the pancreas?

Q14 Comment on the relationship between the development of pancreatitis and alcohol consumption.

Q15 Why do you think a high amylase and lipase concentration was found in Jude's blood?

Q16 Explain why chronic pancreatitis leads to malnutrition and weight loss.

CASE STUDY 43 The producer's stomach ache

Learning outcomes

On completion of this case study, you should be able to:

- describe the symptoms of a peptic ulcer;
- explain how the secretion and motility of the stomach is controlled;
- outline the mechanism of gastric acid secretion and the mode of action of H_2 antagonists and proton pump inhibitors;
- outline the lifestyle advice which is appropriate for peptic ulcer patients.

Part 1

Patterson, a young TV producer, leads a high-pressure life, smoking more or less continuously, eating irregularly and drinking large amounts of strong coffee to keep him going. His alcohol consumption is also high as he often feels the need to 'chill out' with a bottle of wine in the evening. He has little time to cook and many of his evening meals are takeaways. Patterson has suffered from peptic ulcers for several years. His symptoms include acute bouts of central chest pain just below the sternum (epigastric) with nausea and sometimes vomiting. In between these episodes, he suffers from heartburn. His symptoms sometimes wake him at night and are worse when he has eaten late, just before retiring to bed. He takes antacids to control mild symptoms and, when the condition becomes severe, takes a prescribed H_2 antagonist, ranitidine.

As a New Year resolution, Patterson decided to reduce his smoking, alcohol and coffee consumption, realizing that they made his symptoms worse. However, in spite of this, his pain has recently become more frequent and severe. On some nights, Patterson can only sleep if propped up by several pillows.

Q1 What are the functions of the stomach?

Q2 Where do peptic ulcers commonly occur in the gastrointestinal tract?

Q3 Which cells in the stomach secrete HCl and pepsinogen?

Q4 List the components of gastric secretion and the approximate volume of gastric juice secreted per day.

Q5 How is gastric secretion controlled before food enters the stomach and during a meal?

Q6 Describe the normal motility of the stomach and the factors which promote gastric emptying.

Q7 What is heartburn and how might use of extra pillows help Patterson to sleep?

Part 2

Patterson is referred to a gastroenterologist for endoscopy to examine the gastric mucosa. There are signs of inflammation in both the antrum and body of the stomach and an area of ulceration is visualized in the pylorus. Tests for *Helicobacter pylori* (*H. pylori*) are positive. Patterson was treated successfully with a course of antibiotics to eradicate the *H. pylori* infection and a proton pump inhibitor.

Q8 What is a peptic ulcer?

Q9 How does the normal stomach protect itself against digestion by the mixture of HCl and pepsin in gastric juice? How do over-the-counter medicines such as ibuprofen and aspirin and infection with *H. pylori* affect this mechanism?

Q10 Are any pharmacological agents available which can increase protection of the gastric mucosa?

Q11 Why may reduction in alcohol and caffeine help relieve Patterson's symptoms?

Q12 What is the overall aim of the pharmacological treatment of peptic ulcers?

Q13 What is the mechanism of action of ranitidine?

Q14 An alternative drug mechanism to reduce gastric secretion is proton pump inhibition. How do proton pump inhibitors reduce secretion?

Q15 What general advice would be suitable to give to people, like Patterson, who suffer from recurrent ulcer problems?

CASE STUDY 44 Daria's abdominal pain

Learning outcomes

On completion of this case study, you will be able to:

- describe the structure and function of the large intestine;
- outline the control of motility in the colon and rectum;
- review the types and functions of dietary fibre;
- outline the major characteristics of diverticular disease.

Part 1

Daria, a 60-year-old housewife, was taken to hospital this evening with severe abdominal pains. Six hours previously the pain started as a mild cramping sensation in the centre of her abdomen; now it is much more severe, particularly in the lower left part of her abdomen. Until today, Daria had appeared healthy with normal bowel movements. However, she revealed to hospital staff that during the last year she had suffered several episodes of abdominal discomfort and moderate pain lasting a few hours.

Examination showed that her lower abdomen was a little distended and very tender, but there were no other abnormalities. Her blood pressure was low at 102/70 mmHg and her temperature slightly raised at 38.5 °C. A blood sample showed normal haemoglobin and red cell number, but a raised white blood cell count. No intestinal obstruction showed on an X-ray of her abdomen. In view of her symptoms and age, a provisional diagnosis of diverticulitis was made.

Q1 Describe the anatomy of the colon and outline the absorptive function of this structure.

Q2 How does the mucosa of the colon differ from that of the ileum and jejunum?

Q3 How is colonic motility normally controlled?

Q4 The colon contains large numbers of bacteria, which begin to colonize it soon after birth. What are the beneficial actions of colonic bacteria?

Q5 What are diverticula?

Q6 Diverticula occur most often in the sigmoid colon; this area of colon is involved in up to 90% of cases of diverticulitis. What is the anatomical position of the sigmoid colon?

Q7 What components make up the fibre content of the diet in the United Kingdom and North America?

Q8 Diverticulitis usually occurs after the age of 35 years, particularly in Europeans and North Americans, who have a diet which is relatively low in fibre. What is the effect of dietary fibre on the gut and on the transit of gut contents?

Part 2

Daria was given pain relief, intravenous antibiotics and a fluid diet for a few days, and her condition rapidly improved. After four days she was switched to oral antibiotics and a normal diet was gradually introduced. Her recovery was uneventful.

Q9 What features of the history suggest that Daria was suffering from an infection?

Q10 In the general population, which groups of people are likely to be at highest risk of diverticular disease?

Q11 Lack of dietary fibre is associated with constipation. List some other causes of chronic constipation.

Q12 If no treatment is available to patients with diverticulitis and the infection and inflammation continue, what might be the consequence?

Q13 Is diverticulitis likely to be a recurrent condition or is Daria now completely cured of her problem?

Q14 What advice would you give Daria in order to reduce the probability of recurrence?

CASE STUDY 45 That bloated feeling

Learning outcomes

On completion of this case study, you will be able to:

- outline the functions of the small intestinal mucosa;
- review the digestion of carbohydrate, protein and fat and the absorption of the products of digestion in the small intestine;
- outline the characteristic features of malabsorption disorders;
- describe the major features of celiac disease (gluten-sensitive enteropathy).

Part 1

Chloe, a young secretary, visited her doctor because of abdominal bloating and bouts of diarrhoea, which have recently become more frequent. She explains that she first thought her symptoms were due to infection or food poisoning, but now they are so frequent that something else must be causing her problems. Questioning reveals that she has previously experienced several symptoms characteristic of intestinal malabsorption. Between bouts of diarrhoea Chloe's faeces are pale, bulky and malodorous and difficult to flush away. When the diarrhoea occurs, she passes five or six loose, explosive, smelly faeces each day. By evening she is exhausted and is hardly able to climb the stairs to her bedroom.

Chloe's medical notes show that as an infant she was admitted to the children's hospital with suspected celiac disease (gluten-sensitive enteropathy). On discharge, she was prescribed a special diet for four years. Her paediatrician then suggested the gradual introduction of foods previously excluded from her diet, such as bread and breakfast cereals. Since then she has gained weight normally and remained fairly well, except for occasional anaemia. Her blood test now shows microcytic red blood cells; some of her biochemical results are shown below.

	Test	Normal
Hb ($g\,dl^{-1}$)	10.7	12–16
Ferritin ($\mu g\,dl^{-1}$)	0.35	1.5–20
B_{12} ($pmol\,dl^{-1}$)	15.2	11.0–63.0

	Test	Normal
White blood cell count	Normal	
Platelet count	Normal	

Q1 Celiac patients are sensitive to the gluten in wheat, barley and rye, which damages the mucosa of the small intestine. What are the characteristics of the normal mucosa in the duodenum, jejunum and ileum?

Q2 If Chloe is now suffering from celiac disease, there is likely to be a marked reduction in the surface area of her intestinal mucosa. Assuming that her pancreas is normal, what are the likely effects of this condition on protein and carbohydrate absorption?

Q3 What are the likely signs and symptoms of intestinal malabsorption?

Q4 How is Chloe's anaemia related to her celiac disease?

Q5 From the information in the history, from which type of anaemia is Chloe likely to be suffering?

Q6 Describe the absorption of vitamin B_{12} in the intestine. In which part of the intestine does this absorption occur?

Q7 How might Chloe's weakness and fatigue be related to her celiac disease?

Part 2

Biopsy tissue from Chloe's small intestine showed flattened mucosa in all parts because of mucosal atrophy, particularly of the villi. Chloe's small intestinal mucosa was only half as thick as normal. The fat content of her faeces was >8 g per day (normally <6 g daily).

Chloe was prescribed a gluten-free diet and referred to a dietician for dietary help. Celiac patients usually do well on gluten-free diets, but relapse if gluten is reintroduced. After three months on a gluten-free diet Chloe became asymptomatic, more energetic and gained 5 kg in weight. Her haemoglobin and iron stores were still lower than normal so she was prescribed ferrous gluconate, 300 mg daily plus multivitamins including folic acid, vitamin D and calcium.

Q8 What is gluten, and is celiac disease common in Europe?

Q9 How are lipids normally absorbed from the intestine?

Q10 Explain why steatorrhoea is characteristic of malabsorptive conditions, such as celiac disease.

Q11 Some celiac patients suffer from considerable diarrhoea, bloating and flatus. Explain why these symptoms may occur.

Q12 Chloe was prescribed iron (ferrous gluconate) to treat her anaemia after being established on a gluten-free diet. Why is this therapy considered effective at this point and not earlier in her treatment?

Q13 Chloe's treatment included vitamin D and calcium. Why might these be particularly important for a young adult female patient?

Q14 Gluten is present in many home-cooked meals, manufactured foods and ready meals. Restaurant meals and social occasions are difficult for celiac patients as they often cannot tell which foods are gluten-free and suitable for them to eat. Is it likely that Chloe will be able to return to a normal diet in a year or two?

9

Autonomic disorders

CASE STUDY 46 Rob's ocular accident

Learning outcomes

On completion of this case study, you will be able to:

- define *mydriasis* and describe the autonomic control of the pupil;
- describe the factors affecting the diameter of the pupil;
- describe the consequences of an increase in intraocular pressure (IOP) and its pharmacological management.

Part 1

Mature student Rob is 26 years old and is working in a pharmacy research laboratory as part of his work placement module. He deals with many different chemicals on a daily basis. One afternoon while he was getting ready to finish for the day, he noticed that his vision was becoming blurred in the left eye and the laboratory lights were making his eye uncomfortable. After checking his eyes in the mirror, he noticed that the pupil of his left eye was much bigger than the other eye. His left eye was also painful. While he was thinking about visiting a doctor, one of

Clinical Physiology and Pharmacology Farideh Javid and Janice McCurrie

the local doctors walked into the lab. Rob explained his problem and, following questioning, it emerged that Rob had been dealing with atropine, cocaine, morphine and phenylephrine that afternoon. Since it was late in the day, the doctor advised him to go to the local hospital for a proper examination.

Q1 Briefly explain the control of the pupil diameter of the eye.

Q2 Define *mydriasis* and *miosis*, and explain how the diameter of the pupil can be affected by common autonomic agonists and antagonists.

Q3 Under what circumstances could a patient have pupils of an unequal size?

Q4 Could any of the chemicals used by Rob that afternoon have caused his symptoms?

Part 2

During an eye examination at the hospital, it was found that Rob's IOP was above the normal range in the left eye (33 mmHg) and in the right eye was slightly increased (21.5 mmHg). In addition, it was found that the angle between his cornea and iris was very narrow; this was worse in the left eye, where the pupil was dilated. Rob confirmed that eye problems are common in his mum's family.

The doctor made a diagnosis of an acute attack of closed-angle glaucoma, as a result of his narrow drainage angles and the probable exposure to a mydriatic agent.

Q5 What should doctors do immediately for Rob?

Q6 What drugs can be used to lower IOP in this situation?

Q7 What is glaucoma? Comment on its pathophysiology, including the different types of glaucoma.

Q8 What is the normal IOP and how is it maintained?

Part 3

Rob's IOP was successfully lowered with medication, and the ophthalmologist advised him to receive treatment to control IOP until all the presenting symptoms had cleared up, which takes a week or so.

Q9 Comment on the drug treatments for glaucoma by explaining their mechanism of action. Your answer should include some examples of the drugs used.

Q10 Comment on the side effects/contraindications associated with drugs used to treat glaucoma.

Q11 Is there any alternative to drug therapy in treating glaucoma?

CASE STUDY 47 A severe attack of greenfly

Learning outcomes

On completion of this case study, you will be able to:

- describe the anatomical differences between the sympathetic and parasympathetic systems, and the associated neurotransmitter release;
- explain the actions of anticholinesterase enzymes on organs such as the heart, respiratory and central nervous systems, salivary glands, eyes, mucous membrane of the mouth and skeletal neuromuscular junctions;
- review the symptoms of organophosphate toxicity and the use of antidotes.

Part 1

Jim used his redundancy money to start a small business growing pot plants to supply local shops and offices. He had always been a very successful gardener so he was horrified when plants in his new greenhouse suffered a severe attack of greenfly. He collected a large container of commercial insecticide containing the organophosphate malathion from his supplier and set to work with his spray.

After a few minutes of spraying, Jim started to feel very ill indeed and soon collapsed. As a relative novice to commercial gardening, he had not realized that organophosphates are very toxic, as they act as anticholinesterases. He did not appreciate that he should have been using protective clothing when spraying these compounds in a confined space.

Jim's symptoms included severe intestinal cramps, drooling, sweating, lacrimation, agitation, nausea and muscle twitching.

Q1 By which routes could malathion enter Jim's body?

Q2 Which parts of the nervous system appear to have been affected by the insecticide?

Q3 What are the anatomical differences between the sympathetic and parasympathetic divisions of the autonomic nervous system?

Q4 Name the neurotransmitters in the two divisions of the autonomic nervous system.

Q5 Describe the events which lead to the release of transmitter in the parasympathetic nervous system.

Q6 How may anticholinesterases affect neurotransmission within the autonomic nervous system?

Q7 Identify the systems or tissues which appear to have been stimulated to produce the symptoms that Jim experienced and the division of the nervous system which provides innervation to the structures you describe.

Part 2

Fortunately, Jim was found by a colleague soon after collapsing. He was taken into the fresh air and then to the local hospital where his contaminated clothing was removed. He was given breathing support on admission and a drug, pralidoxime, to reactivate his plasma cholinesterase activity. To be fully effective, this drug must be given within a short time of exposure to anticholinesterases, but it can remain active for 24 hours.

Jim was given an 'antidote' to help reduce his symptoms, which were very distressing. After an overnight stay in hospital, he made a good recovery.

Q8 What effects would you expect to observe in (i) the heart, (ii) the bronchi and (iii) the salivary glands following administration of an anticholinesterase?

Q9 Why was breathing support needed in the acute phase of Jim's condition?

Q10 What type of drug could be used as an 'antidote' to relieve the symptoms Jim was experiencing?

Q11 Would the drug you have identified in Question 10 have actions on the skeletal neuromuscular junction? Give reasons for your answer.

Q12 Bethanechol is sometimes used therapeutically to enhance detrusor (bladder) muscle activity when there is evidence of urinary retention. What are the mechanism of action and adverse effects of bethanechol?

Part 3

Some months after Jim's unpleasant experience with the insecticide, he attended an eye clinic for a routine examination. Following the visit, his pupils were widely dilated and he found it difficult to focus on objects.

Q13 What type of drug was likely to have been used for Jim's eye examination to cause these effects?

Q14 If a very large dose of this agent had been instilled into his eye by mistake and had produced systemic actions, what effects would you expect to observe on the heart and on the mucous membranes of the mouth?

10
Reproductive disorders

CASE STUDY 48 Panic of a college girl

Learning outcomes

On completion of this case study, you will be able to:

- describe menstruation and outline its hormonal control;
- review some menstruation-associated problems, their pathology and pharmacological management;
- review the methods of contraception available.

Jane is preparing herself for her final exams at college. She has been working hard to review all the lectures and wants to do her best, since she is planning to go to university to continue her education in medicine. This has really put her under stress and made her lose some weight too. However, in the last few days she has not been able to concentrate very well since she has missed her period. She was too embarrassed to talk to her mother and could not face going for a pregnancy test. She was feeling really down and finally her mum noticed her low mood and sadness. Her mum asked lots of questions and finally Jane gave in and explained about the delay in her menstrual cycle. Her mum tried to calm her down, knowing that Jane had

Clinical Physiology and Pharmacology Farideh Javid and Janice McCurrie

broken up with her boyfriend over two months earlier. However, following further questioning, it emerged that Jane had started a new relationship two weeks earlier, so they decided to book a pregnancy test for the following day. Fortunately, next morning Jane woke up to discover that her menstruation had started. That made her very happy. She was not pregnant after all!

Q1 Define the term *menstruation* and describe the phases of the menstrual cycle.

Q2 Explain the profile of gonadotropic hormones' activity, i.e. luteinizing hormone (LH) and follicle-stimulating hormone (FSH), in a typical 28-day female reproductive cycle

Q3 What is *amenorrhoea*? Comment on its pathophysiology and pharmacological management.

Q4 What is *menorrhagia*? Comment on its pathophysiology and pharmacological management.

Q5 Outline the contraceptive methods available.

Q6 Comment on the composition and mechanism of action of oral contraceptives.

Q7 What risk factors and potential adverse effects should be considered when using combined hormonal contraceptives?

Q8 In case of fertilization, which hormone is initially responsible for interrupting the menstrual cycle? Comment on its source and functions.

Q9 Which hormone is mainly responsible for the contraction of the uterus at birth? From where is it released?

Q10 Why do you think Jane's menstrual cycle was delayed?

CASE STUDY 49 Shabana's monthly problems

Learning outcomes

On completion of the case study, you will be able to:

- review the events and hormonal control of ovulation and the menstrual cycle;
- outline the symptoms and treatment of selected menstrual disorders and syndromes;
- comment on the choice of contraceptive method in older women;
- describe the physiological changes which occur at the menopause;
- review the benefits and disadvantages of hormone-replacement therapy (HRT).

Part 1

Shabana is an overweight 39-year-old woman of small stature who suffered considerable discomfort from menstrual problems as a teenager. She experienced cramping abdominal and back pains followed by nausea and headache, which began just as her period started each month. Her family noted that she became irritable and very easily upset around the time she started to bleed each month. She was prescribed a combined oral contraceptive, which improved her symptoms somewhat. Shabana continued taking the oral contraceptive until her marriage 15 years ago, when she stopped taking the pill. Three years after her marriage she gave birth to a healthy, full-term son.

Q1 What was likely to be the cause of Shabana's teenage period problems?

Q2 Some patients suffer from premenstrual syndrome (premenstrual tension), which is diagnosed if symptoms start before menstruation begins and diminish when the bleeding starts. List the symptoms associated with premenstrual tension.

Q3 Briefly describe the changes in oestrogen and progesterone which occur during the menstrual cycle.

Q4 How are hormones from the hypothalamus involved in the control of the menstrual cycle?

Q5 How does the endometrial lining of the uterus change during the menstrual cycle?

Q6 What is *dysmenorrhoea*? How can it be managed pharmacologically?

Q7 Which hormones are contained in the combined contraceptive pill and how do they reduce the symptoms of dysmenorrhoea?

Part 2

Shabana's periods started again about eight months after the birth of her son. She then began to experience cramping pains at the start of her cycle, and this symptom has become much worse as time passes. Her pains begin a few days before menstruation and continue until two to three days after bleeding has started. Her doctor has recently diagnosed endometriosis. In addition to her menstrual problems, Shabana has gained a substantial amount of weight over the last few years and now weighs 13 st (186 lb).

Q8 Describe the condition of *endometriosis*.

Q9 What treatments may be available to reduce Shabana's monthly problems?

Q10 In younger women, prescription of a combined oral contraceptive would both prevent pregnancy and reduce the symptoms of which Shabana complains. But at Shabana's age the combined pill is not recommended. What factors are considered when deciding to prescribe oral contraceptives in older women and what adverse effects have been linked to the use of these drugs?

Q11 The menopause usually occurs in women between the ages of 45 and 55. What physiological changes and symptoms are characteristic of the menopause?

Q12 Comment on the actions of oestrogen in the body.

Q13 What is HRT? Comment on the use of oestrogen-replacement therapy.

Q14 Comment on the routes of administration of oestrogen.

Q15 Does HRT provide contraception?

Q16 What are the common side effects associated with the use of oestrogens?

Q17 Are there any life-threatening adverse effects of oestrogen therapy?

Q18 Comment on the use of prolonged oestrogen therapy in post-menopausal women.

CASE STUDY 50 Demi's baby

Learning outcomes

On completion of the case study, you will be able to:

- describe the production of gametes in males and females and identify the major causes of male and female infertility;
- list the major functions of testosterone, oestrogen and progesterone;
- explain the hormonal changes occurring in pregnancy and lactation;
- describe the actions of tocolytic and uterine-stimulant drugs;
- outline the composition, production and release of breast milk.

Part 1

Demi and Milton were married five years ago and wanted to start a family immediately. But after nearly three years, Demi had not become pregnant and so decided to consult the family doctor. From her history, it was clear that Demi was a healthy 29-year-old woman who did not smoke and who had no apparent problems with her regular menstrual cycle.

Infertility affects about 15% of couples and can be defined as an inability to conceive following one year of unprotected intercourse with the same partner. Fertility can be reduced by many factors in the male or female partner, or both. In the male, infertility often involves diminished production of sperm or diminished quality of sperm. Since Demi appeared to have a normal menstrual cycle, a sperm count was arranged for Milton. The results showed the count to be within normal limits, but at the lowest end of the range, so an appointment was organized for the couple at a fertility clinic. However, while waiting for the appointment, Demi became pregnant at last.

Q1 Describe the development of the ovum in the ovary and explain how follicular development is controlled.

Q2 What factors might affect the fertility of the female partner?

Q3 Outline the process of spermatogenesis in the testes and the factors which might affect the fertility of the male partner.

Q4 List the functions of testosterone.

Q5 Explain the hormonal changes which maintain pregnancy following the fertilization of the ovum.

Q6 What are the major changes in maternal physiology during pregnancy?

Q7 How does the placenta develop and what are the functions of the placenta during pregnancy?

Part 2

Demi enjoyed a normal pregnancy. During labour, it was necessary to enhance her uterine contractions as the birth was progressing rather slowly. She finally delivered a healthy, full-term baby daughter. Demi decided that breastfeeding would give her daughter the best start in life and, although she did not use any contraceptive method for many months, Demi did not become pregnant again while breastfeeding.

Q8 The process of birth (parturition) involves strong contractions of the uterine muscle (myometrium), which start spontaneously. Which hormones are involved in the initiation and maintenance of uterine contractions in labour?

Q9 Outline the actions of pharmacological agents that can be used to diminish premature uterine contractions, which may occur before gestation is complete.

Q10 What adverse effects on the mother might be induced by intravenous infusion of beta-2-agonists (β_2 agonists), such as salbutamol?

Q11 Which drugs might be useful to enhance and strengthen uterine contractility when labour is not progressing at a satisfactory rate?

Q12 Outline the composition of breast milk and explain the benefits to the baby of breastfeeding.

Q13 How is milk production and the secretion of breast milk controlled?

Q14 Breastfeeding can inhibit ovulation for some months after birth and so can provide a form of contraception. How does lactation inhibit ovulation?

ANSWERS

1

Psychological disorders

CASE STUDY 1 A mother's loss

Q1 Mrs Ford appears to be suffering from depression. Depression is a common psychiatric condition which occurs when sadness or grief is abnormally prolonged and causes dysfunction. It is classified as an *affective disorder*, that is a disorder of mood rather than a disorder involving disturbances of cognition or thought.

Q2 Depression ranges from a mild condition, perhaps associated with a stressful or sad event, such as bereavement, to a severe state which may be accompanied by delusions or hallucinations (psychotic depression). Depressed patients may experience a range of emotional and biological symptoms. They are unhappy, sad and cry for no apparent reason. They are generally negative about life and may be very self-critical, expressing feelings of worthlessness. Patients' energy level appears to be low: they feel constantly tired, lethargic and lack motivation. Patients' sleep may also be disturbed, with characteristic patterns of early waking and inability to return to sleep. Appetite may be reduced or increased and the individual may exhibit psychomotor retardation, a pattern of slow physical movement and response. They may also develop constipation, reduction in libido, anxiety, irritability or tension. Some depressed people show different behaviour patterns and may eat and sleep to excess.

Clinical Physiology and Pharmacology Farideh Javid and Janice McCurrie

Q3 Mrs Ford's symptoms included: feeling constantly down, hopelessness, feeling that life is meaningless, suicidal thoughts, lack of energy and motivation, abandonment of socializing, social activities and other plans, disturbed sleep and eating patterns. These are consistent with the profile of depression.

Q4 The pathophysiology of depression is believed to involve the depletion of noradrenaline (norepinephrine) and serotonin (5-HT) at nerve endings in the brain. These monoamines are important in determining mood.

Q5 There are two main treatments for depression:

(1) drug therapy, for example using antidepressant tablets

(2) talk therapy, such as cognitive behaviour therapy or counselling.

Both of these treatments can be used as a course of therapy over a period of months. They can be used singly or together; the latter will increase the speed of recovery from a period of depression.
In addition, electroconvulsive therapy is available for patients with severe refractory depression. The mechanism by which this treatment alleviates depression is controversial: it may increase the ability of the nerves in the central nervous system (CNS) to respond to noradrenaline and serotonin (5-HT).

Q6

(1) Tricyclic antidepressants, which inhibit or reduce the reabsorption (reuptake) of the main neurotransmitters (noradrenaline and serotonin) into nerve endings.

(2) Monoamine oxidase inhibitors (MAOIs), which were amongst the first antidepressant drugs to be used clinically. They affect one or both of the brain monoamine oxidase enzymes that play a role in the metabolism of serotonin, noradrenaline, dopamine and adrenaline. MAOIs inhibit breakdown of the neurotransmitters important in determining mood, which results in the antidepressant effect.

(3) Selective serotonin reuptake inhibitors (SSRIs), which work by increasing the actions of serotonin at nerve endings. These agents increase the life of serotonin in the synapse and facilitate neurotransmission. The choice of drug is based on the requirement of the individual patient. Any other illness, current drug therapy and previous responses to antidepressants are taken into consideration in choosing an appropriate agent.

Q7 Amitriptyline hydrochloride belongs to the tricyclic group of antidepressant drugs.

Q8 Dosage can be started at 75 mg per day and increased to 150–200 mg per day if necessary. When taken at night, the sedative effect of this agent has a beneficial effect on the patient's sleep pattern.

Q9 Tricyclic antidepressants cause sedation and possess several other side effects. The antimuscarinic (atropine-like) effects of these agents include dry mouth, blurred vision, raised intraocular pressure, postural hypotension, impotence, changes in cardiac rhythm and muscle tremors. They can also cause obstruction of the bladder neck, followed by difficulty in initiating micturition.

Q10 There is a delay of one to two weeks in the onset of response to all antidepressants. This might be due to the time taken to override the feedback mechanisms at the nerve endings. Therefore, Mrs Ford does not need a different medication at this stage, only reassurance that the drug will soon become effective.

Q11 The patient can be prescribed an SSRI as an alternative to amitriptyline.

Q12 An SSRI, such as fluoxetine, can be started at a dose of 20 mg per day. Although similar in their efficacy and time course to the tricyclic drugs, the advantage of the SSRIs is their lack of serious side effects, such as cardiotoxicity, sedation, blurred vision, dry mouth and so on, which are associated with tricyclic antidepressants.

Q13 Patients who take SSRIs might develop gastrointestinal disturbances such as dyspepsia, nausea and vomiting, weight gain, headaches because of the vasodilator effects of serotonin; in some patients insomnia may occur.

Q14 MAOIs, such as phenelzine and isocarboxazid, affect the sympathetic nervous system by inhibiting one or both forms of brain monoamine oxidase. Their sympathomimetic effects can produce a feeling of well-being and increased energy, which is helpful for depressed patients. However, psychosis may occur in a susceptible individual or may follow over-administration of these agents. An increase in sympathomimetic action (such as occurs with use of amphetamines, which increase the release of noradrenaline) can result in a lethal hypertensive crisis. In addition, a hypertensive crisis can also be initiated if the patient consumes a diet rich in amines; foods with a high amine content include cheese, pickles, broad beans and wine.

Key Points

- Depression is characterized by negative, hopeless feelings and feelings of unhappiness for no obvious reason. Patients may feel worthless; mood, sleep and energy levels are affected. Depression disturbs many aspects of our daily life.
- The underlying pathophysiological explanation is believed to be the depletion of noradrenaline and serotonin stores in the body.
- Drug therapy and/or counselling are effective treatments for most patients. Three categories of drugs used as antidepressants are: tricyclic antidepressants, MAOIs and SSRIs.
- Antidepressants have a one- to two-week delay in the onset of therapeutic responses, possibly because of the time taken to override feedback mechanisms at nerve endings.

CASE STUDY 2 A dangerous father?

Q1 The likely diagnosis is mania.

Q2 The symptoms of mania, which is an affective disorder, involve marked elevation of mood. They include: excessive irritability and restlessness with outbursts of anger, elation, enthusiasm and optimism. The patient typically appears overconfident, excessively loud and may make inappropriate demands.

Q3 The underlying pathophysiology of mania is not well understood. It is thought that overstimulation of the noradrenaline transmitter system plays an important role in mania.

Q4 Yes, mania may develop in patients who have been taking antidepressants such as MAOIs or tricyclic antidepressants.

Q5 Lithium (lithium carbonate or citrate) is administered in doses of between 0.2 and 1.5 g daily. The dose is monitored to provide a therapeutic plasma level of 0.4–1.0 $mmol\,l^{-1}$ 12 hours after the most recent dose taken on days 4–7 of treatment. The plasma concentration is then measured every week until the dosage has been stabilized and the required concentration has remained constant for four weeks. Lithium can take several days to become effective. If a patient is suffering an acute attack of mania and is excessively disturbed, treatment with an antipsychotic drug may also be required. The antipsychotic agent can be administered with lithium and the dose is gradually reduced as the lithium takes effect.

Q6 The precise mechanism of action of lithium is not known. It seems to affect (inhibit or block) mechanisms mediated by cyclic adenosine monophosphate (cAMP) and phosphatidylinositol/diacylglycerol secondary messengers. It may inhibit the release of noradrenaline and dopamine. Lithium has the ability to compete with or replace sodium ions in the body, and its excretion is related to sodium levels: if sodium is depleted, lithium is retained and its toxicity increases. Lithium salts take several days (up to a week) to exert a therapeutic effect.

Q7 Lithium has a very narrow therapeutic window (therapeutic index) and overdosage can be fatal. The side effects which patients may experience include gastrointestinal disturbances such as slight nausea and diarrhoea. Anorexia may also occur. In low concentrations lithium induces excessive thirst (polydipsia), which could be due to its effects on sodium retention and inhibitory actions on antidiuretic hormone. There may be mild CNS disturbances, including tremor, sleepiness, dizziness, tinnitus, unsteadiness and blurred vision; patients' speech and cognitive ability may also be affected. Higher concentrations of lithium can cause muscle twitching, convulsions and possibly coma and death. Taking lithium in long-term therapy (over three to five years) damages cells of the nephron. It also affects the cardiovascular system, and overdosage may lead to fatal hypotension.

Q8 The excretion of lithium is reduced in patients taking non-steroidal anti-inflammatory drugs. This leads to increased plasma concentration and enhancement of the effects of lithium. Since lithium is more toxic when sodium is depleted, the use of diuretics, particularly the thiazides, during lithium treatment is contraindicated. High doses of antipsychotic drugs, such as haloperidol, may also be hazardous if used with lithium because they increase neurotoxicity. Some agents used for cardiovascular disorders, for example angiotensin-converting enzyme inhibitors and digoxin, may also increase the neurotoxicity of lithium salts.

Q9 Alternative drugs used to treat mania include: benzodiazepines, carbamazepine or antipsychotic agents such as fluphenazine and risperidone.

Key Points

- Mania is associated with excessive irritability and outbursts of anger, elation, enthusiasm and optimism, which may be due to the overstimulation of the noradrenaline transmitter system.
- Lithium is the first-line drug therapy which might act by inhibiting the release of noradrenaline and dopamine. Lithium has the ability to compete with or replace sodium ions in the body, and its excretion is related to sodium levels: if body sodium is depleted, lithium is retained.
- Lithium has a narrow therapeutic index. A plasma level higher than $1.5\,mmol\,l^{-1}$ causes serious problems and can be fatal.

CASE STUDY 3 Continual concerns for Mr Watson

Q1 The swings of mood from depression to mania suggest a diagnosis of manic depressive disorder (bipolar affective disorder). In this condition, the cycle of manic and depressive periods can take place over months or years, but may occur rapidly over weeks or days; this varies between patients. On the other hand, there may be several episodes of depression which follow each other, or the patient may experience several episodes of mania in succession.

Q2 The precise pathophysiology is unknown. An elevated level of choline in the basal ganglia of patients with mood swings has been suggested. There may also be a change in the metabolism of phospholipids and abnormal energy production in the frontal and temporal lobes. An abnormal metabolism of high-energy phosphates in the brain's frontal lobes has been suggested. There might be a change/reduction in the conduction of impulses along the neurones and/or in neuronal communication pathways.

Q3 Manic depressive disorder involves mood swings, and several agents can be used to stabilize mood: lithium carbonate or citrate are often used. There appears to be little difference in the therapeutic usefulness of these two salts. Lithium salts are widely used mood stabilizers. They are useful prophylactically in treating both acute mania and bipolar conditions.

Q4 A dose of 450 mg is usually given twice a day, and the dose is adjusted to maintain a serum level of 0.5–1.5 mmol l^{-1}. Once the condition is under control, a maintenance dosage is given to obtain a serum level of 0.5–1.0 mmol l^{-1}. Before and during long-term treatment, the patient's renal and thyroid function requires monitoring as kidney damage and hypothyroidism may occur with long-term use.
Because of the serious risks involved in long-term lithium treatment, patients' plasma levels are reassessed regularly, usually every three months. If plasma lithium concentration becomes too high, administration of the drug is suspended and large amounts of sodium salts and fluids are given. Since lithium toxicity is enhanced by sodium depletion, the increased plasma sodium and fluids can reduce its toxic effects.

Q5 Carbamazepine or valproic acid can be used in treating bipolar disorder and are useful for patients who are unresponsive to lithium. Initially, carbamazepine may be given in a divided dose of 400 mg daily. The normal dosage range can increase to 600 mg daily in divided doses, although a maximum dose of 1600 mg daily may be needed in some patients. The initial dosage for valproic acid is 750 mg daily in two or three divided doses, increasing to 1–2 g daily if necessary.

Q6 It is significant that Mr Watson's father also suffered from mood swings, because a patient has an increased risk (approximately 10-fold) of suffering from manic depressive illness if a first-degree relative is similarly affected.

Q7 If a patient is receiving lithium, a lithium treatment card is available to inform the patient how to take the medication, what to do if a dose is missed, when blood tests will be necessary and so on. Patients should be advised to drink plenty of fluids each day and avoid changes in their diet that could increase or decrease their usual salt intake. Patients should not take antidepressants on any sustained basis, as their use may promote mania.

Key Points

- Manic depressive disorder (bipolar affective disorder) is characterized by swings of mood from depression to mania. A patient may be at an increased risk of developing the condition if a first-degree relative is similarly affected.
- The condition may be associated with:
  - an elevated level of choline in the basal ganglia.
  - changes in the metabolism of phospholipids and abnormal energy production in the frontal and temporal lobes.
  - disruption of neuronal communications in the brain.
- Lithium and carbamazepine are used in treating manic depressive disorder.
- Patients on long-term treatment with lithium should be reassessed regularly to avoid the renal and thyroid toxicity that can occur with this agent.

CASE STUDY 4 A scary presentation

Q1 Jo is showing symptoms of anxiety. Many symptoms of anxiety are observed in our fear response to unpleasant or threatening stimuli. In anxiety states these symptoms occur independently of the usual fear-provoking stimuli. So an anxiety disorder is a condition in which a state of anxiety persists without any obvious reason.

Q2 Symptoms of anxiety can include: breathlessness, palpitations (increased awareness of the heart beat, or an irregular heart rhythm), dry mouth, difficulty in swallowing, flatulence, nausea, diarrhoea, tachycardia, dizziness, blurred vision, sleep disturbance, sweating, tension, irritability, restlessness, apprehension, depression, worry, fear. Some patients also report chest pain or chest constriction.

Q3 Somatic symptoms presented in this case are: dry mouth, tachycardia and sweating. Psychological symptoms are: tension, apprehension, irritability, restlessness and difficulty in concentrating. The symptoms usually result from overactivity in part of the autonomic nervous system or increased tension in skeletal muscles.

Q4 The neurotransmitters GABA (gamma-aminobutyric acid) and serotonin (5-HT) are mainly associated with anxiety disorders. In addition the sympathetic component of responses observed in anxiety, which stimulates a dry mouth, tachycardia, sweating and so on, involves the neurotransmitter noradrenaline.

Q5 Jo's tachycardia is due to the activation of the sympathetic nervous system to prepare the body for 'fight or flight'. Stimulation of sympathetic nerves supplying the heart releases noradrenaline, which increases both the rate and force of cardiac muscle contraction via beta-1-receptor (β_1-receptor) activation.

Q6 Other conditions that could be confused with anxiety include endocrine disorders such as thyroid problems and hypoglycaemia, autonomic disorders, drug/alcohol misuse and other CNS disorders such as panic disorder.

Q7 Treatment of anxiety disorders involves the use of anxiolytic preparations such as benzodiazepines. The 'fight or flight' symptoms can be controlled by sympathetic β-adrenoceptor antagonists, such as propranolol. Non-pharmacological behavioural therapy is also successfully used in the treatment of anxiety disorders.

Q8 Anxiolytics are a group of drugs that reduce the symptoms of anxiety mentioned earlier. They are among the most frequently prescribed drugs and can be divided into two subgroups: benzodiazepines and non-benzodiazepines.

Q9 Benzodiazepines (BZDs), such as diazepam or alprazolam, act on neuronal benzodiazepine receptors (located adjacent to GABA receptors) in the CNS. Stimulation of these receptors leads to increased inhibition at postsynaptic neurones mediated by GABA. This inhibition results in depression in the limbic and subcortical areas of the brain. Benzodiazepines thus cause sedation and muscle relaxation.

Diazepam can also be used as a muscle relaxant and has anticonvulsant activity when given intravenously. Alprazolam possesses antidepressant properties in addition to its anxiolytic actions.

Q10 A major problem associated with benzodiazepines is the development of tolerance, a gradual increase in the dose needed to elicit the therapeutic effect and dependence in chronic use. Following the cessation of treatment, the patient may suffer from rebound anxiety and insomnia. Withdrawal from benzodiazepines also occasionally causes bizarre visual disturbances.

Q11 Non-benzodiazepines, such as buspirone, may be used to treat anxiety. Buspirone is an agonist at 5-HT_{1A} receptors. It has been suggested to act by inhibiting the neuronal firing via these receptors which in turn reduces serotonin turnover in the CNS. The anxiolytic action of this agent may take days or weeks to develop, but there are less troublesome side effects. The exact mechanism of action of buspirone in reducing anxiety is not yet known. The usual daily dosage is from 15 to 30 mg, used in divided dosages.

Q12 Yes, anxiety could develop into a phobic state. A *phobic state* is defined as anxiety/fear triggered by a single stimulus or set of stimuli that would not normally be of concern. A panic disorder involves sudden and unpredictable episodes of acute anxiety, with feelings of fear and terror, usually accompanied by severe physical symptoms. The tendency to panic disorder may be genetically transmitted. A change in levels of lactic acid or carbon dioxide in the blood may play a part in this disorder.

Key Points

- Anxiety disorder is a condition in which a state of anxiety persists without any obvious reason.
- The symptoms usually result from overactivity in part of the autonomic nervous system or increased tension in skeletal muscles.
- Somatic symptoms of anxiety are: breathlessness, palpitations (increased awareness of the heart beat, or an irregular heart rhythm), tachycardia, dry mouth, difficulty in swallowing, flatulence, nausea, diarrhoea, dizziness, blurred vision, sleep disturbance, sweating.
- Psychological symptoms are: tension, irritability, restlessness, apprehension, depression, worry, fear.
- The neurotransmitters GABA and serotonin (5-HT) are mainly associated with anxiety disorders. In addition the sympathetic component of responses is mediated by noradrenaline.
- Treatment includes: non-pharmacological therapies, such as psychological approaches, and pharmacological treatment, use of anxiolytic preparations, such as benzodiazepines, non-benzodiazepines, such as buspirone, and β-adrenoceptor antagonists such as propranolol.

CASE STUDY 5 Fussy Jane

Q1 The likely diagnosis is obsessive–compulsive disorder (OCD).

Q2 Patients with OCDs are unable to stop thinking certain thoughts and undertaking particular actions. In some patients the disorder is associated with anxiety and/or depression. The person with OCD is conscious of the uselessness of their thoughts and actions; however, he or she is unable to stop the cycle, and this can cause great distress. Patients may develop obsessions for a ritual of repetitive cleaning, counting or checking: if interrupted during their activity, they need to start again from the beginning of the ritual. These repetitions greatly interfere with the patient's normal lifestyle and the daily life of their family. The behaviours may go on for many years and are then often quite resistant to treatment.

Q3 An increase in the level of glucose metabolism has been suggested as the underlying pathophysiology of this condition in certain brain areas, such as frontal lobes, caudate nuclei and cingulated gyri in patients suffering from OCP. In addition, the involvement of serotonin has been suggested in washing, cleaning and danger-avoidance behaviours.

Q4 (A) OCD is associated with changes in serotonin (5-HT) metabolism, and therefore the use of the antidepressant agents will be useful. For example, clomipramine (a tricyclic antidepressant) may be prescribed: 100–150 mg daily, starting with 25 mg per day and then increasing the dose over two weeks. (This is a larger dose than that used for depression.) Also, SSRIs, such as fluoxetine (20–60 mg daily) or fluvoxamine (50–200 mg daily), can be prescribed.

Q4 (B)These drugs act by blocking serotonin reuptake at synapses, thus increasing the level of serotonin at the synaptic junction. In increase in the level of serotonin may diminish the repetitive behaviours.

Q5 Some patients are helped by behavioural therapies, and anxiolytic drugs have been found to provide short-term relief of the symptoms.

Key Points

- In OCD patients are unable to stop thinking certain thoughts and undertaking particular actions. Their repetitive behaviour greatly interferes with the patient's normal lifestyle and the daily life of their family.
- Cortical regions of the brain have been suggested to be involved in the mediation of the symptoms. Both serotonin and an increase in the level of glucose metabolism may play a role in this disorder.
- Treatments include behavioural therapies, anxiolytic drugs and antidepressants such as the SSRIs, fluoxetine or fluvoxamine.
- SSRIs increase serotonin concentration at synapses, which may be responsible for diminishing repetitive behaviours.

CASE STUDY 6 David's withdrawal

Q1 The likely diagnosis is schizophrenia. This is the most common form of psychosis.

Q2 The positive symptoms are symptoms such as hallucinations, which are usually auditory, and delusions. Some thought disorders and abnormal behaviours may also be placed in this category.

Q3 The negative symptoms are features such as social withdrawal, apathy and lack of purposeful behaviour. There is usually a reduction or flattening of emotional responses.
Patients also develop cognitive disruption so that speech and written communication is affected, that is they may use a string of words with no rational meaning. Many schizophrenic patients describe religious experiences, such as hearing the word of God, or claim particular artistic sensitivity, for example they understand the hidden meaning of poems, novels, pictures and so on.

Q4 Positive and negative symptoms usually occur together; thus, these patients withdraw from society and cannot maintain relationships. Commonly, they have persecuted feelings, for example that somebody is following or checking up on them.

Q5 Positive: delusions, auditory hallucinations; negative: social withdrawal and neglect.

Q6 Yes, see answers presented above.

Q7 Other conditions which may present similar symptoms include: drug-induced psychosis such as one brought on by lysergic acid diethylamide, or LSD, or amphetamine, personality disorder or affective psychosis. In older patients dementia may present with schizophrenia-like symptoms, but these patients usually have significant memory deficits, which do not occur in schizophrenia.

Q8 Dopamine is thought to be the main neurotransmitter associated with schizophrenia. But there is evidence of the involvement of other neurotransmitter systems, particularly glutamate, and also serotonin (5-HT) and GABA.

Q9 Schizophrenia appears to involve both genetic and environmental factors. Possible causes of schizophrenia are:

(1) an increase in the release of dopamine from the nerve terminal.

(2) the development of a hypersensitivity in dopamine receptors.

(3) problems with inactivation of dopamine at the synapse.

(4) failure of dopaminergic feedback mechanisms.

(5) development of an imbalance between the activity of the dopamine and glutamate systems.

Q10 Haloperidol is an antipsychotic or neuroleptic agent. It is an antagonist at dopamine receptors, particularly of the D_2 subtype. These drugs help to control the symptoms (mainly the positive symptoms) of schizophrenia by antagonizing the dopamine receptors in different brain areas, such as the frontal and temporal lobes. Antipsychotic agents, such as haloperidol, take days or weeks to achieve their therapeutic effect and may produce some motor disturbances.

Q11 Other neuroleptic agents include phenothiazines, such as chlorpromazine, promazin and thioridazine, and thioxanthines, such as flupenthixol. The non-specific blockade of dopaminergic receptors afforded by these drugs leads to development of side effects, such as endocrine dysfunction and extrapyramidal motor symptoms. The unwanted antagonism of motor tracts results in extrapyramidal side effects, such as Parkinsonism and tardive dyskinesia. The latter is associated with involuntary movements of the face, limbs and trunk. Chronic neuroleptic therapy can inhibit the release of GABA. This in turn leads to changes in mobility.
The agents used in treating schizophrenia are most successful in treating the positive symptoms; negative symptoms, such as apathy and social withdrawal, seem to be less responsive to current drug treatment.

Key Points

- Schizophrenia is the most common form of psychosis and appears to involve both genetic and environmental factors.
- The symptoms may be classified as positive and negative.
- Positive symptoms are symptoms such as hallucinations, which are usually auditory, and delusions; some thought disorders and abnormal behaviours may also be placed in this category.
- The negative symptoms are features such as social withdrawal, apathy and lack of purposeful behaviour, reduction or flattening of emotional responses and development of cognitive disruption so that speech and written communication are affected.

- Neurotransmitters associated with schizophrenia are dopamine as the main neurotransmitter, plus other neurotransmitter systems, particularly glutamate, but also serotonin (5-HT) and GABA.
- Antipsychotics are also referred to as neuroleptic drugs, such as haloperidol, phenothiazines (e.g. chlorpromazine, promazin and thioridazine) and thioxanthines (e.g. flupenthixol). All theses agents are used in treating the symptoms and are mostly successful with positive symptoms.
- Haloperidol helps to control the positive symptoms of schizophrenia by antagonising dopamine receptors (D_2 receptors) in several brain areas, but may produce motor disturbances.
- Other neuroleptic agents such as phenothiazines, which are non-selective dopamine antagonists, can cause endocrine dysfunction as well as extrapyramidal side effects.

CASE STUDY 7 Forgetful mum

Q1 The likely diagnosis is Alzheimer's disease. In Britain, Alzheimer's disease must be diagnosed and treatment initiated in a specialist clinic.

Q2 Alzheimer's disease is a dementia associated with a progressive loss of cognitive function. Some loss of intellectual ability with age is normal and the rate at which it occurs is very variable. In Alzheimer's disease this loss of cognitive function is pronounced. It is associated with cortical neurodegeneration, and can occur in mid-adult life or later in the absence of any other form of brain insult, such as drug toxicity, stroke or head injury. The symptoms start with short-term memory loss. It is more common in later life: approximately 5% of 65-year-olds and 30% of 85-year-olds suffer from Alzheimer's disease.

Q3 Post-mortem examinations of the brains of patients with Alzheimer's disease show loss of cortical neurones and abnormal depositions of proteins in the cerebral tissues. The normal structure of the brain is modified by β-amyloid plaques, sometimes called *senile plaques*, and neurofibrillary tangles produced by abnormal neurones. Neurochemical changes in the brain occur, mainly involving cholinergic systems but also other neurotransmitters and neuromodulators.

Q4 Cholinergic nerves are mainly affected. There is reduction in the enzyme choline acetyltransferase and a deficit in acetylcholine.

Q5 Donepezil is a reversible inhibitor of acetylcholinesterase, which is administered once daily. The starting dosage is 5 mg daily at bedtime, increasing if necessary after one month to 10 mg daily.

Q6 Cholinesterase inhibitors block the action of the enzyme acetylcholinesterase (which normally hydrolyses acetylcholine) and so terminate its activity. These drugs increase the life of released acetylcholine at the synapse, leading to an enhancement of acetylcholine activity. Drug treatment for Alzheimer's disease is supervised in specialist clinics where the patient's cognitive function can be assessed at approximately three-monthly intervals. About half the patients treated show a decreased rate of cognitive decline while receiving this type of drug.

Q7 The adverse effects associated with cholinesterase inhibitors are related to excessive cholinergic stimulation. These include gastrointestinal disturbances such as abdominal cramps and nausea, salivation, sweating, flushing, bronchoconstriction and urinary incontinence.

Q8 The drugs currently licensed for Alzheimer's disease in Britain are cholinesterase inhibitors, with one exception – memantine hydrochloride. This agent is a *N*-methyl *D*-aspartate (NMDA) receptor antagonist that reduces glutamate transmission. Side effects of dizziness, confusion, tiredness and hallucinations have been reported.

Key Points

- Alzheimer's disease is a dementia associated with a progressive loss of cognitive function.
- There is a loss of cortical neurones and abnormal deposition of proteins β-amyloid and neurofibrillary tangles in the cerebral tissue.
- Cholinergic nerves are mainly affected.
- Cholinesterase inhibitors, such as donepezil, are the first line of drug treatment as they can decrease the rate of cognitive decline in some patients. But there are side effects with these agents that are related to excessive cholinergic stimulation, for example abdominal cramps, bronchoconstriction, salivation and so on.
- Memantine hydrochloride, an NMDA receptor antagonist, reduces glutamate transmission and has shown beneficial effects.

CASE STUDY 8 Disruptive John

Q1 The doctor's diagnosis was attention deficit hyperactivity disorder (ADHD).

Q2 ADHD is a neurodevelopmental disorder that affects children's development. The disorder tends to run in families. It is characterized by inattention, impulsiveness, uncooperativeness, aggressiveness and hyperactivity. The behaviour is not appropriate for the age of the child. There are three subtypes: hyperactive/impulsive, inattentive and combined. Many children with this disorder are impulsive and disruptive, cannot sit quietly and continually seek attention.

Q3 The actual cause of ADHD is not known. However, many factors such as brain damage, genetic predisposition, reduction in dopamine levels, encephalitis, food hypersensitivity and high levels of environmental lead have been suggested as contributing to the development of ADHD.

Q4 ADD is attention deficit disorder. Patients with ADD do not show hyperactivity.

Q5 Methylphenidate (for example Ritalin) is a CNS stimulant. Treatment can be started at a dose of 5 mg per day and this can be increased by 5 mg every two days. The maximum daily dose should not be more than 60 mg. The last dose should be given four hours before bedtime. This drug is not recommended for children under the age of six years. Treatment of ADHD in Britain is normally initiated in a specialist clinic, after which it may be continued by family doctors.

Q6 The exact mechanism of action of methylphenidate is not known. However, CNS stimulants generally cause the release of neurotransmitters such as serotonin, dopamine and noradrenaline. The drug is normally used as part of a comprehensive treatment programme for ADHD under specialist supervision.

Q7 Side effects of methylphenidate are: appetite suppression, nausea, abdominal pain, nervousness, irritability and insomnia. The patient's blood pressure needs to be checked as use of the drug may involve headaches and dizziness. In the long term, the medication may affect a child's height and weight and his growth should be monitored during prolonged treatment. The effectiveness of the medication should also be reassessed before the onset of puberty.

Q8 Yes, individuals with ADHD in childhood may need psychological counselling throughout their lives. ADHD may manifest itself in different ways in adults since they may find it easier to cope with the condition. However, affected adults may have problems in concentrating on a task, despite their efforts to focus.

Key Points

- ADHD is a neuro-developmental disorder that affects children's behaviour and development. The behaviours are not appropriate for the age of the child. Patients with ADD do not show hyperactivity.
- There are three subtypes: hyperactive/impulsive, inattentive and combined.
- Factors such as brain damage, genetic predisposition, reduction in dopamine levels, encephalitis, food hypersensitivity and high levels of environmental lead have been suggested to contribute to the development of ADHD.
- First line of treatment is the use of methylphenidate or Ritalin.
- Psychological counselling may be needed in conjunction with drug treatment.

2

Neurological disorders

CASE STUDY 9 Mrs Smith's tremor

Q1 The most likely diagnosis for these symptoms is Parkinson's disease. Parkinson's disease is a progressive movement disorder and is the leading cause of neurological disease in people over 65 years of age.

Q2 Prominent symptoms of the condition include: tremor, rigidity and hypo- or bradykinesia. Typically, patients first notice symptoms, like hand tremor or foot dragging, on one side of the body. The symptoms subsequently spread to both sides. Because muscle rigidity increases and there is difficulty in starting

Clinical Physiology and Pharmacology Farideh Javid and Janice McCurrie

and stopping voluntary movement, patients whose disease is at an advanced stage have a shuffling type of walk. Yes, it can be hereditary but this is relatively uncommon; the condition is believed to be mainly due to the effects of environmental factors.

Q3 The pathophysiology of Parkinson's disease is related to deficiency of the neurotransmitter dopamine. There is damage/degeneration of the dopaminergic pathways in the nigrostriatal area, which consists of the substantia nigra together with fibres synapsing in the caudate, putamen and basal ganglia. In these associated areas, the neuronal dopamine stores might also be depleted.

Q4 In Parkinson's disease the diminished influence of dopamine on the excitatory actions of acetylcholine in the basal ganglia leads to an imbalance in favour of the cholinergic effects. This, in turn, results in the symptoms of Parkinson's disease. Thus, the focus of drug therapy could be on balancing the dopaminergic and cholinergic activity either by reducing cholinergic function or by enhancing dopaminergic actions. Different treatments can successfully provide some alleviation of the symptoms but they do not prevent progression of the disease. Treatment may consist of:

(1) Replacing dopamine that is lost in the pathway using levodopa (L-dopa).

(2) Use of dopamine agonists to activate existing dopamine receptors.

(3) Use of monoamine oxidase B inhibitors (MAOB) to lengthen the duration of action of dopamine, by inhibiting its metabolism.

(4) Use of dopamine-releasing agents, such as amantadine.

(5) Use of COMT (catechol-O-methyltransferase) inhibitors, such as tolcapone, which leads to an increase in the transport of L-dopa into the brain.

(6) Use of antimuscarinic drugs.

Q5 The rationale for the use of L-dopa is to replace dopamine that has been lost in the pathway. L-dopa, an amino acid precursor of dopamine, can be given, since, unlike dopamine, it can cross the blood–brain barrier. However, most of the administered L-dopa is decarboxylated in the peripheral tissues of the liver and gut, only 10% of L-dopa passes into the brain. The peripheral actions of this agent lead to unpleasant side effects of nausea, vomiting, anorexia and postural hypotension. To increase the amount of L-dopa in the brain and reduce its decarboxylation in the periphery, an extracerebral dopa-carboxylase inhibitor, such as carbidopa, is added to the treatment. This inhibitor increases availability of L-dopa, enabling an increased quantity of the drug to enter the brain. For example, carbidopa (or benserazide) is given in combination with L-dopa a, which prevents the peripheral metabolism of L-dopa to dopamine, leading to an increase in the level of L-dopa in the brain. L-dopa, usually in combination with carbidopa, is regarded as a first-line treatment for Parkinson's disease.

Q6 Dopamine is unable to cross the blood–brain barrier and so cannot enter the brain. A precursor of dopamine, L-dopa, which can penetrate the blood–brain barrier, is given instead (see above).

Q7 The action of amantadine is to enhance dopamine transmission in the central nervous system (CNS) by facilitating the release of dopamine from central neurones. Additionally, it may also increase the synthesis of dopamine and inhibit dopamine reuptake mechanisms. Amantadine may also induce an anticholinergic effect. It can be given in the early stages of Parkinson's disease, when tremor is not prominent, or used in combination with L-dopa at the more advanced stages of the disease. Used alone, it is generally regarded as having only a modest antiparkinsonian effect. This agent was presumably judged to be unsuitable to treat Mrs Smith's symptoms of tremor, stiffness and difficulty in moving up or down stairs.

Q8 Acetylcholine is normally in balance with dopamine in the basal ganglia, but in Parkinson's disease dopamine levels are reduced and the effects of acetylcholine become more pronounced. To restore the balance, antimuscarinic agents are used to antagonize the excitatory actions of acetylcholine. They seem to be effective for tremor and reduce the secretion of saliva, digestive juices and sweat.

Q9 Antimuscarinic drugs are generally used in younger patients and not in the elderly. This is because elderly patients may be suffering from urinary retention or closed-angle glaucoma; these drugs are contraindicated in patients with the above problems. The side effects of antimuscarinic agents, such as dry mouth, tachycardia, dizziness and constipation, are troublesome to most patients, but more so in the elderly. Antimuscarinics also interact with many other drugs that the elderly are likely to take, such as tricyclic antidepressants, antihistamines and sublingual nitrates. Antiparkinsonian drugs may cause confusion in elderly patients. It is usual to start treatment with the lowest effective dose, increasing the dosage slowly if required.

Q10 Final confirmation of the diagnosis can only be made at post-mortem examination.

Q11 The major consideration in selecting the most appropriate medication for Parkinson's disease is the severity of the symptoms and the age of the patient. Treatment for this condition should be started under the supervision of a specialist doctor.

Key Points

- Parkinson's disease is a progressive movement disorder and is the leading cause of neurological disease in the elderly.
- Symptoms include: tremor, rigidity and hypo- or bradykinesia. Patients first notice symptoms, like hand tremor or foot dragging, on one side of the body and these eventually spread to both sides. Patients whose disease is at an advanced stage have a shuffling type of walk.
- The disease can be hereditary and is believed to be mainly due to the effects of environmental factors.
- The underlying pathophysiology is related to dopamine deficiency, which results in the imbalance of cholinergic and dopaminergic activities.
- Treatment may consist of using L-dopa, dopamine agonists, MAOB inhibitors, dopamine-releasing agents, COMT inhibitors and antimuscarinic drugs.
- Younger patients may be prescribed antimuscarinic drugs. These are not as suitable for elderly patients. Elderly patients are more likely to be also suffering from glaucoma or urinary retention, for which the drugs are contraindicated.

CASE STUDY 10 Rose's loss of consciousness

Q1 Epilepsy.

Q2 Epilepsy is a condition which is characterized by recurrent seizures. A seizure is an intense, sudden uncontrolled burst of abnormal neuronal activity across the cerebral cortex of the brain.

Q3 A possible cause is an abnormality of the inhibitory action of gamma-aminobutyric acid (GABA) transmission in the brain. Failure of the GABA transmitter to exert its inhibitory action on a burst of high-frequency firing of CNS neurones leads to the involvement of normal neurones in spreading the abnormal electrical activity. The release of excitatory glutamate may also be implicated in the pathophysiology of seizures.

Q4 An electroencephalogram (EEG) is a recording of electrical activity arising from the cortical surface of the brain. It is recorded from scalp electrodes on 16 channels simultaneously. The technique is non-invasive and is not painful. The main uses of electroencephalography are to investigate sleep and its disorders and to diagnose epilepsy. The wave/spike patterns produced can be analysed to reveal alterations in or to localize areas of the specific electrical activity associated with seizures. The EEG can also be used medico-legally to determine whether a person is actually 'brain dead'.

Q5 The two main categories of epileptic seizures are: partial and generalized. Partial seizures are characterized by a burst of abnormal activity in a localized area of the brain. In generalized seizures the abnormal electrical activity involves the whole brain and always results in loss of consciousness.

Q6 The symptoms of partial seizures may consist of repetitive contractions of a single group of muscles, or abnormal sensations such as hearing voices or seeing coloured lights. Some partial seizures result in abnormal behaviour, such as purposeless hand-rubbing, alterations of mood or behaviour resembling drunkenness. Generalized seizures may result in repeated muscle contractions throughout the whole body (grand mal or tonic clonic epilepsy) or may consist of a sudden loss of consciousness for a short period (petit mal epilepsy or absence seizure).

Q7 Generalized seizure, petit mal epilepsy.

Q8 The management of epilepsy is by surgery or drug therapy. In drug therapy anticonvulsant or anti-epileptic drugs are given that work via two mechanisms: first, they induce a stabilizing effect on excited neurones by reducing sodium ion exchange across the cell membrane, so preventing the spread of neuronal excitation. The second mechanism is by reducing the focus of neuronal discharge by increasing the activity of GABA, which acts by antagonizing synaptic transmission. This action reduces or abolishes the excessive electrical discharge.

Q9 Sodium valproate increases the GABA content of the brain. It is thought to achieve this by initiating the release of GABA at synapses and weakly inhibiting GABA transaminase, an enzyme which inactivates GABA. There may also be some effect post-synaptically on sodium channels that enhances GABA action. The side effects of this agent include gastrointestinal disturbances, tremor, transient alopecia and increased appetite with weight gain. However, the main problem associated with use of valproate is potential hepatotoxicity. Hepatotoxicity is a rare, but serious, side effect of treatment with valproate, so monitoring liver function is advised for the first six months of therapy.

Q10 Examples of three drugs used to treat epileptic seizures are:

(1) Sodium phenytoin, which mediates the removal of sodium ions from intracellular space during the refractory period of an action potential.

(2) Lamotrigine, which exerts its effects by reducing the exchange of sodium ions across the cell membrane. In addition it reduces the release of excitatory glutamate, which may be implicated in the pathophysiology of seizures.

(3) Vigabatrin, which acts by irreversibly inhibiting the enzyme responsible for the metabolism or degradation of GABA.

Part 2

Q11 Yes, valproate can cause irregular menstruation. Both the condition of epilepsy and the treatment of epilepsy can alter fertility. Oligomenorrhoea (irregular, long menstrual cycles >32 days) or amenorrhoea (absence of menstruation for three months) may occur, but many of the affected women still ovulate.

Q12 Yes, valproate increases the risk of giving birth to a baby with spina bifida or anencephaly since valproate antagonizes the effects of folic acid and causes these neural tube defects. Thus women who suffer from epilepsy and who are taking valproate should receive a much higher dosage of folic acid (5 mg daily) compared to healthy women, who are advised to take 400 μg daily.

Q13 Although the incidence of congenital abnormalities is somewhat increased in women treated for epilepsy, Rose has a >90% chance of delivering a completely healthy baby while taking valproate. Stopping the medication is not recommended since seizures can harm the developing baby. There is evidence that minor fits have no effect on the developing baby, but major seizures early in pregnancy are associated with major malformations.

Also if the fit results in a fall, there may be injury to the foetus or a miscarriage could occur. In addition, sudden withdrawal of the drug could cause status epilepticus, seizures which follow one another without return of consciousness. This is a medical emergency and can result in serious consequences for both mother and child. Therefore, Rose should continue with her medication under supervision.

Q14 An alternative drug associated with a lower incidence of congenital abnormalities, which would reduce the risk of spina bifida or anencephaly in Rose's baby, is lamotrigine. If this agent were prescribed, it would be given in addition to the valproate, starting at a very low dose and slowly increasing the dosage while reducing valproate dosage. In this way lamotrigine would gradually replace the valproate.

Key Points

- Epilepsy is characterized by recurrent seizures, which are intense, sudden uncontrolled bursts of abnormal neuronal activity across the cerebral cortex of the brain.
- The abnormality in the inhibitory action of GABA transmission in the brain and/or the release of excitatory glutamate may be implicated in the pathophysiology of seizures.
- The two main categories of epileptic seizures are: partial and generalized.
- Partial seizures are bursts of abnormal activity in a localized area of the brain and cause symptoms such as repetitive contractions of a single group of muscles or abnormal sensation, such as hearing voices or seeing coloured lights, and abnormal behaviour.
- Generalized seizures are bursts of abnormal electrical activity in the whole brain and result in loss of consciousness (petit mal epilepsy or absence seizure) or may result in repeated muscle contractions throughout the whole body (grand mal or tonic clonic epilepsy).
- Drug therapy includes the use of anticonvulsant or anti-epileptic drugs, such as sodium valproate, sodium phenytoin, lamotrigine, vigabatrin.
- Sodium valproate increases the GABA content of the brain. Side effects of this agent include gastrointestinal disturbances, tremor, weight gain, irregular menstruation and, potentially, hepatotoxicity. Monitoring liver toxicity is advised for the first months of therapy.

CASE STUDY 11 Another day away from the office

Q1 Sue appears to be suffering from migraine.

Q2 The symptoms of migraine include: severe and recurrent headache lasting between two hours and several days, visual and gastrointestinal disturbances (including photophobia, nausea and vomiting) and a sensation of 'pins and needles'.

Q3 Migraine affects approximately 10% of the population; the incidence is higher (double) in women compared to men. The first attack may occur at any time between childhood and early adulthood. Approximately 20% of women in their early forties suffer from migraine.

Q4 The most common causes of migraine are dietary and stress factors. Food and drinks which contain high levels of tyramine can initiate migraine. Examples of food and drinks containing a high concentration of tyramine are cheeses, particularly blue cheeses, chocolate, wine, caffeine, alcohol, tea, citrus fruits, eggs and fried and spicy foods. In the 'stress' category: anxiety, tension, sleeplessness and a change in the normal daily routine can also trigger migraine. In addition, other environmental factors, such as flashing lights, can act as triggers. Changes in the level of female hormones (oestrogen and progesterone) during pregnancy and menstruation and in response to oral contraception and hormone-replacement therapy can promote migraine.

Q5 There are two main types of migraine:

(1) Migraine without aura (visual disturbances), which is the common type of migraine from which the majority of people suffer. The headache is severe and is usually accompanied by photophobia, nausea, vomiting and prostration, which may last for many hours.

(2) Migraine with aura, which is the classic type of migraine. The aura is followed approximately 30 minutes later by severe, throbbing headache and its sequelae. This type affects 20% of people who suffer from migraine.

Q6 The symptoms of aura can consist of various visual disturbances, usually including a flickering pattern and often followed by blind spots. This disturbance may be associated with a wave of spreading depression in cortical neurones, perhaps triggered by emotional or biochemical changes. Other sensory disturbances, such as 'pins and needles', may occur. These sensations move up one arm to the face and may last from 10 to 60 min. Some individuals experience other symptoms, such as mood swings, hyperactivity, hunger, thirst or cravings. In addition, some individuals may develop fluid retention, oliguria or diuresis.

Q7 The mechanisms which trigger migraine remain controversial. The underlying pathophysiology could be due to vasoconstriction of the cerebral arteries, causing transient ischaemia. This would be followed by compensatory vasodilation of the cerebral blood vessels to protect the ischaemic areas. This vasodilation may lead to an increase in intracranial pressure, which causes a severe headache. These events may be followed by changes in nerve activity and neurotransmitter levels. Inflammatory components are also likely to be involved in the pathology of this condition.

Q8 Serotonin (5-HT) has been implicated in the pathogenesis of migraine. It has been shown that levels of serotonin increase just before a migraine attack, and fall sharply at the beginning of the headache. Furthermore, the cerebral blood flow is reduced during the early and aura phases, and then increases during the headache phase. Serotonin affects vascular tone, and many of the drugs which are effective in treating migraine are either serotonin agonists or antagonists.

Q9 Yes. The aura, severe headache, nausea and vomiting are consistent with the classic type of migraine.

Q10 Agents that are used to treat an acute migraine attack include analgesics, such as aspirin, opioids, ergot derivatives and triptans (5-HT_1 agonists). In some people a simple analgesic such as soluble aspirin or a non-steroidal anti-inflammatory drug (NSAID) is satisfactory. However, intestinal peristalsis is reduced during a migraine attack, so drug absorption may be inadequate. Triptans, such as sumatriptan, are very effective and may be used during the headache phase.

If migraine attacks become very frequent or disabling, preventive treatment should be discussed. Patients are counselled to consider possible trigger factors, such as stress, disturbances of normal sleep patterns, recent adoption of irregular lifestyle, or food or alcohol triggers. The following types of drug can be used for the prevention of migraines: β-adrenoceptor antagonists, serotonin antagonists and tricyclic antidepressants. Calcium channel blockers are used in the prophylaxis of cluster headaches. The latter are more common in males than females, unlike migraine. Cluster headache pain is very intense and steady rather than throbbing. Cluster headaches tend to occur frequently over a period of days and are then followed by a headache-free period of weeks or months.

Q11 Sumatriptan is a serotonin agonist (5-HT_1 agonist) that reduces the cerebral vasodilation which is believed to cause migraine headache.

Key Points

- Migraine is associated with severe and recurrent headache lasting between two hours and several days, with visual and gastrointestinal disturbances, photophobia, nausea and vomiting, and a sensation of 'pins and needles'.
- The most common causes of migraine are dietary and stress factors.
- Migraine can occur without aura (visual disturbances) or with aura.
- A compensatory vasodilation of the cerebral blood vessels may lead to an increase in intracranial pressure, which causes a severe headache. Serotonin (5-HT) has been implicated in the pathogenesis of migraine.
- Drug treatments include analgesics such as aspirin, opioids, ergot derivatives and triptans (5-HT_1 agonists).
- For prevention of migraines, the following types of drug can be used: β-adrenoceptor antagonists, serotonin antagonists and tricyclic antidepressants.

CASE STUDY 12 Drooping eyelids

Part 1

Q1 Myasthenia gravis.

Q2 Patients with myasthenia gravis suffer from extreme muscle weakness and fatigue, particularly after repeated muscle contraction. A noticeable feature of myasthenia gravis is that the upper eyelids droop (ptosis) because of the unconscious and repeated use of the muscles involved in blinking, and these eyelid muscles show fatigue and weakness before any other skeletal muscle is affected. Weakness of other muscles innervated by the cranial nerves is usually also visible early, resulting in a loss of the person's normal facial expression. Their vision is affected, their jaw may drop and their speech may become slurred. The condition occurs mostly in women, with a peak incidence in the third decade.

Q3 Myasthenia gravis affects the function of the junction between motor nerves and skeletal muscle. It is an autoimmune disease in which antibodies are formed against the acetylcholine receptor proteins on muscle membranes. The antibodies attack the acetylcholine receptors at the skeletal neuromuscular junction and therefore acetylcholine fails to bind to them. This results in muscle weakness, particularly of the eye, lips, tongue, throat, neck and shoulders. Movement will be limited when the limb muscles are affected, making any repetitive action, for example in lifting, walking, running and climbing stairs, difficult to sustain.

Q4 Yes. Mrs Downs is reporting general muscle weakness and fatigue, drooping eyelids, which is characteristic of this condition, and some difficulty in focusing. Myasthenia will affect the ocular muscles of nearly all patients, leading to double vision. In addition, Mrs Downs has experienced forearm weakness as she reported problems in carrying heavy shopping. Abnormal fatigability in limb muscles makes it difficult for patients to lift and carry objects and possibly even to lift the arm to comb their hair. Leg muscles are generally affected later.

Q5 A synapse is the contact point between a neurone and another cell (e.g. a second neurone or a muscle cell). For an electrical signal to pass from one neurone to another it must cross a synapse.

Q6 There are two different types of synapses:

(1) Electrical synapses, which are tubular structures (called *connexions*) and form gap junctions: the membranes of the two cells are separated by a distance of 2 nm. They may allow the two-way transmission of impulses;

transmission is fast. They are rarely found in the CNS, but are more likely in cardiac or smooth muscle cells.

(2) Chemical synapses, for example between two neurones or between a neurone and a muscle fibre; transmission is slower since there is a delay of 0.5 ms because of a gap of 20 nm between the cells. For transmission to occur the chemical transmitter must be made and stored in vesicles at the presynaptic side. The transmitter is ready to be released whenever an action potential arrives at the presynaptic nerve. Because the transmitter is only on one side of the synapse, the impulse can move in only one direction.

Q7 The arrival of the action potential depolarizes the nerve terminal. Voltage gated calcium channels open, which leads to an increase in the level of calcium in the nerve terminal, causing vesicles containing the transmitter to merge with the synaptic membrane of the terminal. Exocytosis occurs, which in turn releases transmitter into the synaptic cleft. Depending on the type of transmitter and synapse, the transmitter can then bind to a receptor located on the postsynaptic membrane, be broken down by hydroxylase enzymes or can be taken up again into the presynaptic terminal for recycling.

Q8 The number of synapses decreases with age. There are 10^{16} synapses in childhood and the number decreases to 10^{15} in old age.

Q9 Neostigmine, which is an anticholinesterase agent, inhibits the action of cholinesterase enzymes on acetylcholine. This results in an increase in the level, and prolongs the action, of acetylcholine at the synapse, which enhances neuromuscular transmission and muscle strength.

Q10 Anticholinesterase agents have muscarinic side effects. They produce effects similar to muscarinic stimulation, such as increased salivation, sweating, gastric secretion and gastrointestinal upsets, increased intestinal motility and diarrhoea, an increase in bronchial secretions and muscle twitching (fasciculation). The administration of an antagonist such as atropine will reduce the incidence of these side effects, since atropine has specific muscarinic antagonist activity. An alternative agent, pyridostigmine, may be used if neostigmine produces intolerable side effects. Pyridostigmine is an anticholinesterase with a slower but longer duration of action. It has comparatively mild gastrointestinal side effects, but an antimuscarinic drug should also be given with this agent.

Part 2

Q11 The number of normal acetylcholine receptors decreases as the disease progresses. This reduces the effectiveness of anticholinesterases. In such cases, immunosuppressant therapy, using a corticosteroid, can be used. This will help to reduce the formation of antibodies to acetylcholine receptors. In addition, removal of the thymus gland improves the clinical condition of many patients.

Key Points

- Myasthenia gravis affects the function of the junction between motor nerves and skeletal muscle.
- It is an autoimmune disease in which antibodies attack the acetylcholine receptors at the skeletal neuromuscular junction and therefore acetylcholine fails to bind to them. The condition results in muscle weakness, particularly of the eye, lips, tongue, throat, neck and shoulders.
- Movement will be limited when the limb muscles are affected, making any repetitive action, for example in lifting, walking, running and climbing stairs, difficult to sustain. Vision will also be affected.
- Drug treatments include the use of anticholinesterase agents, such as neostigmine and pyridostigmine, in conjunction with an antimuscarinic drug such as atropine.

3
Endocrine disorders

CASE STUDY 13 An agitated mother

Q1 The thyroid gland is situated in front of the larynx in the neck. It contains two lobes joined with a narrow central region called the *isthmus*. The gland consists of round follicles, lined by rings of cuboid epithelium. The follicles store a sticky colloidal material, thyroglobulin. Thyroid hormone is derived from this colloid. The secretion of free T_3 and T_4 into the bloodstream occurs when thyroid stimulating hormone (TSH) stimulates the proteolytic digestion of thyroglobulin.

Q2 Thyroxine (T_4), the major hormone secreted together with triiodothyronine (T_3). T_3 is more active metabolically than T_4. In addition the hormone calcitonin is secreted by the medullary cells of the thyroid gland.

Q3 The following diagram shows the control of thyroid hormone secretion:

Secretion of T_3 and T_4 is controlled by circulating TSH released from the anterior pituitary gland. When levels of T_3 and T_4 rise, the secretion of TSH is reduced. TSH secretion is also controlled by thyrotrophin releasing hormone (TRH) from the hypothalamus.

Clinical Physiology and Pharmacology Farideh Javid and Janice McCurrie

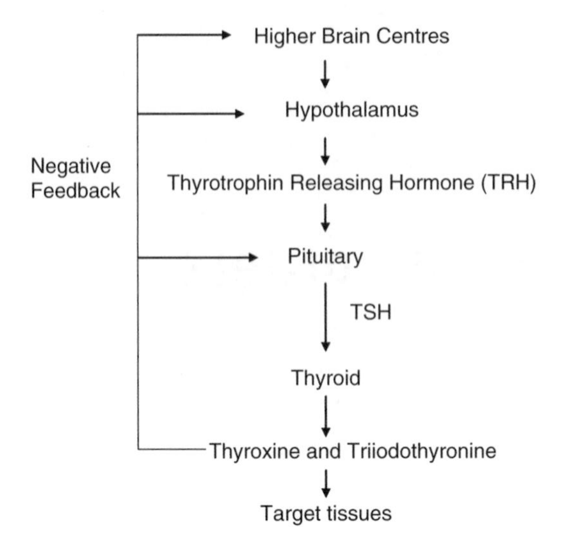

Q4 In hyperthyroidism there is too much thyroid hormone, triiodothyronine (T_3) and thyroxine (T_4), in the body, which raises the metabolic rate of all tissues. Symptoms of hyperthyroidism (thyrotoxicosis) include: an enlargement of the thyroid gland (goitre), increased metabolism, tachycardia and cardiac arrhythmia, excessive sweating, diarrhoea, nervousness, agitation, dyspnoea, tremor, fatigue, muscular weakness, anxiety and weight loss despite increased appetite and staring, protruding eyes (exophthalmus). In addition, osteoporosis and amenorrhoea may also occur. Although there is a drive to increase activity, the person has little energy reserve and tires quickly.
Mrs Kay does not show all the symptoms described above, but yes, her nervousness and agitation, fatigue, sweating, tachycardia, weight loss and changed reflexes are consistent with a profile of hyperthyroidism.

Q5 Graves' disease.

Q6 Graves' disease is an autoimmune disease caused by the presence of thyroid stimulating antibodies which attack the TSH receptors in the thyroid gland, preventing the TSH from binding to its receptors.

Q7 In addition to Graves' disease, the overactivity of one or more nodules in the thyroid can cause toxic multinodular goitre. An acute inflammation of the thyroid gland can also lead to thyroiditis, which produces a transient thyrotoxicosis.

Q8 The drug treatment of thyrotoxicosis involves using antithyroid drugs: carbimazole (which is converted to the active compound methimazole) and propylthiouracil inhibit the synthesis of thyroid hormone. Propylthiouracil also inhibits peripheral conversion of T_4 to T_3. Many of the symptoms of hyperthyroidism can also be alleviated by β-adrenoceptor antagonists. Iodine prevents the release of thyroid hormone and reduces the vascularity of the thyroid gland. This is not a very effective treatment and is used mainly before surgery on the gland.

Q9 The main serious side effects associated with the use of carbimazole are bone marrow depression, neutropenia and agranulocytosis. So patients should be asked to report any sign of infection, especially sore throat, mouth ulcers, high temperature and rashes, since they are signs of bone marrow depression.

Key Points

- Hyperthyroidism is associated with an increase in the levels of thyroid hormone, triiodothyronine (T_3) and thyroxin (T_4), in the body, which in turn raises the metabolic rate of all tissues.
- Symptoms of hyperthyroidism (thyrotoxicosis) include: an enlargement of the thyroid gland (goitre), increased metabolism, tachycardia and cardiac arrhythmias, excessive sweating, diarrhoea, nervousness, agitation, dyspnoea, tremor, fatigue, muscular weakness, anxiety and weight loss despite increased appetite, and staring, protruding eyes (exophthalmus). In addition, osteoporosis and amenorrhoea may also occur. Although there is a drive to increase activity, the person has little energy reserve and tires quickly.
- Drug treatment of thyrotoxicosis involves using antithyroid drugs: carbimazole, propylthiouracil, beta-adrenoceptor antagonists and iodine.

CASE STUDY 14 A vague and sleepy lady

Part 1

Q1 Thyroid secretion is controlled by two feedback loops. Secretion of T_3 and T_4 is stimulated by TSH from the anterior pituitary gland. The secretion of TSH is controlled by the hypothalamus via production of TRH. TRH is secreted by the hypothalamus into the hypophyseal portal blood flow. Stimuli such as a very cold environment influence the secretion of thyroid hormones by affecting the hypothalamus and increasing the release of TRH.

Q2 In addition to poor memory, factors which suggest a diagnosis of hypothyroidism include: cold intolerance, cold extremities, slowed reflexes, low resting heart rate, slow thought processes, depression and sleepiness/lack of energy, appetite suppression associated with weight gain and raised blood lipids, which may lead to increased atherosclerosis.
Hypothyroidism is a relatively common endocrine abnormality with a prevalence in the United Kingdom of 1.4% in women and $<0.1\%$ in men. The incidence increases with age. When the deficiency of thyroid hormones is serious or long-standing, patients suffer skin thickening and may lose body hair. This condition is known as *myxoedema* since a mucopolysaccharide accumulates in the subcutaneous tissues producing a non-pitting oedema, which gives a puffy appearance to the skin, particularly noticed on the face. Mild cases of hypothyroidism are difficult to differentiate from the changes seen in normal ageing. Depression is a relatively common condition in hypothyroid elderly patients, who may suffer delusions or appear demented. In severe cases myxoedema patients may develop a greatly diminished level of consciousness known as *myxoedema coma*.

Q3 The thyroid hormones thyroxine and triiodothyronine have many metabolic effects. In adults they increase metabolic rate, oxygen and calorie consumption, stimulate carbohydrate metabolism and turnover of protein, deplete fat stores and increase catabolism of free fatty acids. Thyroid hormones stimulate heart rate and force and increase pulmonary ventilation, gastrointestinal motility and central nervous system (CNS) activity. Actions on the heart can result in an increased incidence of dysrhythmias. Thyroid hormones are critical for the normal growth and development of the infant, particularly in respect of skeletal growth and maturation of the CNS.

Q4 Weight gain without an increase in appetite is a feature of myxoedema, a severe form of thyroid deficiency. Some of the factors which appear to be involved include: reduced metabolic rate, and oxygen and calorie consumption. In addition complexes of protein with polysaccharides and other substances accumulate under the patient's skin to promote water absorption and retention and increase body weight. This process causes the puffiness of skin observed in myxoedema. When the patient is successfully treated, these complexes are mobilized and there is a diuresis.

Q5 Prominent symptoms of thyroid hormone deficiency are: lack of energy, lethargy, low metabolic rate, slow thinking and speech, poor memory, intolerance to cold, bradycardia and weight gain. In the infant, mental impairment and retardation of growth occurs, leading to the condition of cretinism. Hypothyroid states are treated with oral levothyroxine sodium. The starting dose for elderly people is usually 50 μg daily; this is increased in steps of 50 μg until the patient's metabolism is normalized.

Q6 Common causes of hypothyroidism include autoimmune conditions such as Hashimoto's thyroiditis, in which an immune reaction to thyroid tissue or to thyroglobulin develops, resulting in deficiency of thyroid hormone production. Another cause is iodine deficiency. In order to continue producing T_3 and T_4, the thyroid gland is continuously stimulated and gradually enlarges, sometimes producing a large swelling or goitre. Development of this condition may involve ingestion of antithyroid substances in the diet or in medicines, or may be due to simple iodine deficiency in the diet.

Part 2

Q7 Both hypothyroidism and anaemia can cause fatigue, deficits in concentration and sleepiness. However, Zadie's haematocrit (red cell mass) is within the normal range as is her haemoglobin concentration. If Zadie were suffering from anaemia, her haemoglobin would be low and the haematocrit would be reduced.

Q8 Zadie's blood pressure and heart rate are rather low for an elderly lady, but they are probably within the normal range for a fit person. However, Zadie is clearly not fit!

Q9 Secretion of T_3 and T_4 is normally stimulated by TSH, released from the anterior pituitary. A rise in circulating thyroid hormone concentration reduces the production of TSH by negative feedback. If the gland fails to produce adequate thyroid hormone, production of TSH is not inhibited and its secretion continues to increase. Patients with hypothyroidism generally have reduced T_3 and T_4 production and raised plasma TSH, which is seen in Zadie's case.

Q10 Many endocrine secretions are controlled by negative feedback systems. When the thyroid is stimulated and thyroid hormone concentration increases, it inhibits production of TSH to reduce further stimulation of the gland. As thyroid hormone secretion then diminishes, the negative feedback on the anterior pituitary is reduced and TSH secretion increases again. Basically, in negative feedback loops a rise in hormone production decreases the release of its stimulating hormone, and vice versa.

Q11 Hypothyroid patients typically have increased levels of blood lipids. Hypercholesterolaemia increases atherosclerosis and so increases risk of myocardial infarction and stroke so treatment of hypercholesterolaemia is desirable. Lowering cholesterol has been shown to reduce the progress of atherosclerosis and the risk of cardiovascular disease. The agents available include: cholesterol binding resins, which complex cholesterol in the gut to prevent reabsorption, for example colestyramine, and fibrates, such as gemfibrozil, which markedly reduce the circulating very low-density lipoprotein (VLDL) concentration in plasma. The statins, such as simvastatin, which competitively inhibit 3-hydroxy-3-methylglutaryl coenzyme A and so decrease the synthesis of cholesterol, have been shown in many studies to improve outcome for patients at high risk of cardiovascular disease.

Key Points

- Hypothyroidism is characterized by decreased plasma concentration of thyroid hormones, T_3 and T_4, with increased concentration of TSH. Severe thyroid hormone deficiency is known as *myxoedema*.
- Symptoms of hypothyroidism include lack of energy, fatigue, sleepiness, poor memory, lack of concentration, bradycardia and cold intolerance. Women are more often affected than men and the condition is more common in the elderly.
- The treatment is oral levothyroxine, starting at 50 μg for elderly patients
- The hypothyroid condition is associated with increased blood lipid levels, which can be treated using statins or cholesterol binding resins such as colestyramine.

CASE STUDY 15 A dehydrated business woman

Q1 The likely diagnosis for Nazira is hyperparathyroidism.

Q2 The hormones that are normally involved in the control of calcium balance are parathyroid hormone (PTH) from the parathyroid gland; calcitonin, which is secreted by the thyroid gland; and 1,25-dihydroxycholecalciferol (1,25-DHCC, or calcitriol), which is produced in the kidneys. Calcitonin reduces the level of plasma calcium by attenuating its release from bone and by increasing its excretion. The PTH and 1,25-DHCC increase the level of plasma calcium by two mechanisms: (1) a combination of an increase in calcium absorption by the gut and an increase in the release of calcium from bone and (2) a reduction in both bone formation and calcium excretion. The three hormones act together to maintain the physiological level of calcium and normal bone turnover. Over 95% of body calcium is located in bone as hydroxyapatite.

Q3 The parathyroid glands are four structures usually located on the dorsal surface of the thyroid gland, two on the right and two on the left. In most individuals the glands are embedded in thyroid gland tissue; however, in some individuals, they are separate from the thyroid gland.

Q4 The parathyroids produce a peptide hormone, PTH, which controls the level of calcium in the body. A sensor on the surface of the parathyroid cells monitors blood calcium concentration and PTH is secreted in response to a fall in plasma calcium ion concentration. An increase in the level of PTH leads to hypercalcaemia (raised blood calcium); conversely, a reduction in the level of PTH leads to hypocalcaemia. PTH acts on the kidney to reduce reabsorption of phosphate and at the same time to increase reabsorption of calcium. In addition, it promotes the release of calcium and phosphate into the blood by activating osteoclasts, which break down the inorganic matrix of bone. PTH also increases the absorption of calcium by the mucosal cells of the intestine. The latter is a rather slow, indirect action mediated by PTH stimulation of calcitriol secretion by the kidney.

Q5 An increase in the secretion of PTH results in development of hyperparathyroidism. This causes hypercalcaemia and bone demineralization. It can promote the following effects on the body:

(1) An overstimulation of osteoclastic activity in bone, which increases release of calcium from the matrix into the bloodstream. Continuation of this process leads to bone deformities.

(2) An increase in the level of calcium delivered to the kidneys leads to hypercalciuria and kidney stones. Proximal tubular functions of the

nephron are disturbed by excess PTH, leading to production of alkaline urine; since extra calcium is filtered into the renal tubules while phosphate reabsorption is decreased, the formation of renal stones is facilitated. The production of alkaline urine, with associated metabolic acidosis, is due to the reduced proximal reabsorption of bicarbonate and its increased elimination in the urine.

(3) Hypercalcaemia can also lead to hypophosphataemia, which has deleterious effects on the cardiovascular, respiratory and muscular systems, leading to general debility.

Q6 The cause of excessive PTH secretion may be primary, secondary, or tertiary. In primary hyperparathyroidism one or more glands show exaggerated functions, do not respond to the normal feedback via serum calcium and secrete PTH autonomously. However, the most common cause (80% of the cases) is a benign tumour of parathyroid tissue in one of the glands. Secondary hyperparathyroidism is due to the development of hypocalcaemia. There is an increase in the level of PTH; however, the kidneys, which are major target organs for this hormone, fail to respond and therefore the level of calcium remains low. In tertiary hyperparathyroidism, which occurs in chronic renal failure, the hyperplastic parathyroid cells lose their sensitivity to circulating calcium levels. This leads to autonomous secretion of PTH.

Q7 The total serum calcium concentration is normally about 9.5 mg dl^{-1}. Approximately half of this is bound to plasma protein, mostly to albumin. Most of the remainder is unbound or ionized calcium, which is the physiologically and clinically important form. *Hypercalcaemia*, normally defined as a serum concentration of >12 mg dl^{-1}, may sometimes be caused by excessive consumption of calcium in the diet. More important pathologically is malignant disease. Hypercalcaemia occurs when there are bone metastases associated with breast or prostate cancer. However, many tumours can produce a PTH-like protein causing elevated serum calcium levels. Furthermore, intoxication and immobilization of vitamin D or excess vitamin D may also cause hypercalcaemia.

Q8 Many of the symptoms of hypercalcaemia are non-specific. In excitable cells the membrane potential is stabilized (hyperpolarization) and the cells become less excitable; fatigue, weakness, lethargy, confusion, anorexia, nausea and constipation are common. There are changes in the electrocardiogram (ECG), leading to heart block and other cardiac rhythm disturbances. A condition similar to diabetes insipidus also occurs with symptoms of polydipsia and polyuria. These symptoms are due to a reduction in the responsiveness of the renal tubules to antidiuretic hormone (ADH).

Q9 The hypercalcaemia which occurs in hyperparathyroidism may be reduced by administration of a loop diuretic such as furosemide, which helps calcium excretion. Bisphosphonates, which prevent bone resorption and so reduce calcium release from bone, can be used to treat hypercalcaemia associated with malignancies. Calcitonin may also be useful in treating the hypercalcaemia associated with cancer, as it reduces calcium levels both by attenuating its renal reabsorption and by increasing calcium deposition in bone.

Q10 Hypoparathyroidism is a rare condition which could develop following a decrease or deficiency of PTH secretion, or be due to reduction in the effectiveness of PTH on target cells. Deficiency of PTH may follow damage to the parathyroid glands during thyroid surgery. Deficiency of PTH decreases the concentration of plasma calcium and increases phosphate concentration. A metabolic crisis can occur when hypocalcaemia coincides with hypoparathyroidism.

Q11 A reduction in or lack of PTH activity reduces the level of plasma calcium because of decreased reabsorption of calcium from the intestine, renal tubules and bone. At the same time, because of the renal retention of phosphates, the levels of serum phosphate will be increased (hyperphosphataemia).

Q12 Hypocalcaemia increases the excitability of nerve and muscle cells. It is associated with reduction of the threshold potential necessary for initiation of nerve impulses, and consequently cell excitation occurs following a slight stimulus. The resulting symptoms include prolonged muscle spasms, which can particularly affect the face and limbs, hyper-reflexia, clonic-tonic convulsions and laryngeal spasm, which could cause asphyxia.

Q13 Magnesium is a major intracellular cation which acts as a co-factor in many intracellular enzyme reactions. Plasma concentration is normally $2\,mg\ dl^{-1}$. This ion is abundant in the diet, and hypomagnesaemia is relatively uncommon, unless there is malabsorption or excessive loss via the kidney. However, when present, hypomagnesaemia can lead to hypoparathyroidism. Adjustment to the levels of magnesium can shift the function of the parathyroid glands back to normal. Chronic alcoholism, malnutrition, malabsorption, renal tubular dysfunction and excessive use of diuretics, such as loop and thiazide diuretics, may lead to hypomagnesaemia. Symptoms of magnesium deficiency include depression, confusion, muscle weakness and sometimes convulsions.

Q14 Calcium gluconate and vitamin D (D_3) can be used. In the presence of bile salts, oral vitamin D can be used to treat hypocalcaemia since it initiates the absorption of calcium from the diet by the intestinal mucosa. However, high levels of vitamin D can lead to calcification of the tissues, which causes severe muscular weakness and abdominal pain.

Key Points

- Hyperparathyroidism results from an increase in the secretion of PTH.
- The most common cause (80% of the cases) is a benign tumour of parathyroid tissue in one of the glands.
- Hyperparathyroidism leads to hypercalcaemia and bone demineralization.
- The resultant hypercalcaemia can also lead to hypophosphataemia, which has deleterious effects on the cardiovascular, respiratory and muscular systems, leading to general debility.
- Drug treatments include the use of a loop diuretic such as furosemide, biphosphonates and calcitonin.

CASE STUDY 16 Brian's weight gain

Q1 Each adrenal gland is composed of an outer cortex and an inner medulla. The cortex consists of three layers where several steroid hormones, synthesized from cholesterol, are produced and secreted. The outer layer of the cortex, the zona glomerulosa, produces the mineralocorticoid aldosterone. The zona fasciculata lies under this layer and, together with the inner layer, the zona reticularis, secretes glucocorticoids, mainly cortisol, corticosterone and androgens.
The adrenal glands are located in the retroperitoneal space on top of each kidney. The name *adrenal* indicates their proximity to the kidneys.

Q2 The adrenal cortex secretes three main groups of hormones: mineralocorticoid (e.g. aldosterone), glucocorticoids (e.g. cortisol, cortisone and corticosterone) and gonadocorticoids (mainly androgens with small amounts of progesterone and oestrogen). Aldosterone is a salt-retaining hormone which acts on the kidney to facilitate sodium and water reabsorption.
Glucocorticoids alter the DNA transcription process in their target tissues. They adapt the body to stress. Glucocorticoids also promote the rapid provision of glucose and fatty acids for energy and permit the responses of the circulation required in reaction to stressful events. When glucocorticoid secretion is chronically increased, there is suppression of the immune system: patients experience poor wound healing and increased susceptibility to infection.

Q3 The adrenal medulla is quite distinct from the adrenal cortex in both origin and structure, forming an endocrine component of the autonomic nervous system (ANS). The medullary cells, chromaffin cells, have similar properties to postganglionic neurons of the ANS. Chromaffin cells synthesize epinephrine (adrenaline) and norepinephrine (noradrenaline) from the amino acid tyrosine. Stimulation of the adrenal medulla via sympathetic preganglionic nerves leads to the release of epinephrine and norepinephrine and prepares the body for 'fight or flight'. Heart rate, blood pressure and skeletal muscle blood flow is increased and metabolism is stimulated. Blood flow to the skin and gastrointestinal tract is decreased, so diverting blood to the skeletal muscles and heart.

Q4 Glucocorticoid secretion is controlled by the hypothalamus and anterior pituitary gland. Corticotrophin releasing factor (CRF) is produced in the hypothalamus and travels in the hypophyseal portal blood vessels to the anterior pituitary to release ACTH (adrenocorticotrophic hormone). There is a daily (circadian) rhythm in CRF and ACTH secretion, with a peak in the morning between 7 and 9 a.m. and a low point during the night.

A major stimulus to CRF and ACTH release is stress, which could, for example, be the result of accident or trauma, infections, anxiety or reduction in blood sugar. ACTH release is regulated by CRF and blood levels of ACTH regulate cortisol secretion via a negative feedback mechanism. As cortisol levels rise, ACTH secretion is reduced, and vice versa.

Q5 Symptoms and signs of Cushing's syndrome include:

- Accumulation of fat tissue in the abdomen and behind the shoulders (buffalo hump).
- Thinning of legs and arms because of catabolic effects of cortisol on muscle and adipose tissue. Muscle wasting, weakness and difficulty in movement may occur as a result of the protein catabolism.
- Protein loss also induces osteoporosis, which can result in bone fractures.
- Loss of collagen results in thinning of the skin, and so the capillaries become more visible. Small vessels can rupture and cause bruising.
- Increase in urination with resulting dehydration.
- Mood changes ranging from euphoria to depression may occur.
- Suppression of the immune system, because of long-term increase in the level of cortisol.
- Pigmentation of skin, because of stimulation of melanocytes by ACTH.
- Increase in the level of androgens may cause amenorrhoea in female patients.

Addison's disease is relatively uncommon and is caused by reduced secretion of all the hormones produced by the adrenal cortex.
Symptoms and signs of Addison's disease include:

- hypoglycaemia, leading to lethargy and weakness
- weight loss
- hypovolaemia and hyperkalaemia
- anorexia, nausea and vomiting
- hyperpigmentation, for example on hands or face, or vitiligo (patches of de-pigmented skin)
- in severe cases, hypotension and cardiovascular collapse.

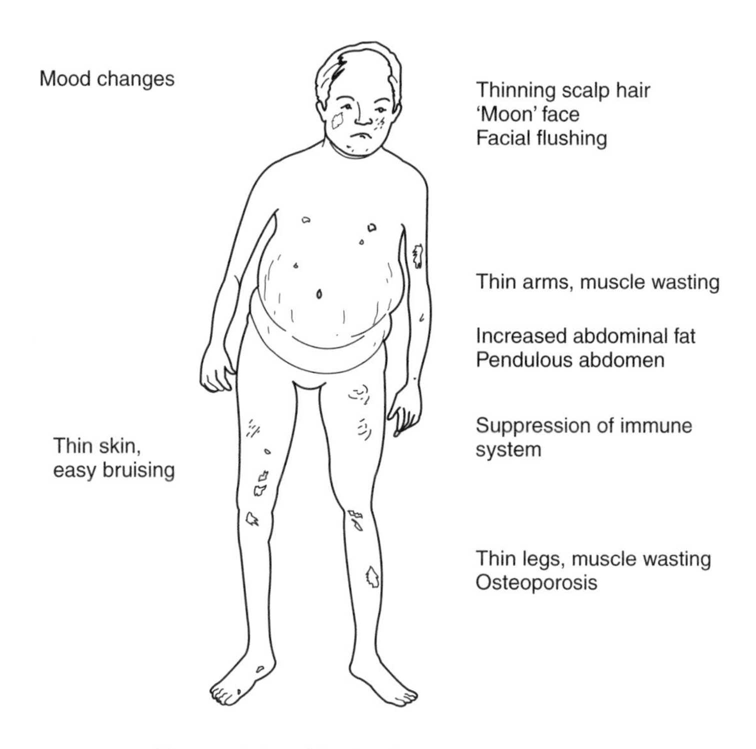

Characteristics of Cushing`s disease

Q6 Brian's muscle weakness, tiredness, weight gain, redistribution of body fat (truncal obesity), limb muscle atrophy, frequent infections, changes in mood and attitude are characteristic of Cushing's disease, which is caused by hypersecretion of glucocorticoids. If his plasma and urinary cortisol is also found to be elevated above the normal value or there is found to be overproduction of ACTH, this confirms the diagnosis. Cushing's syndrome is less common, exhibits more severe symptoms and occurs when there is excessive secretion of cortisol.

Q7 Increase in the level of circulating cortisol leads to Cushing's disease. A high level of plasma cortisol may occur following long-term steroid medication. Tumours of the adrenal cortex and ectopic ACTH, secreted by oat cell carcinomas of the lung, also increase the plasma concentration of cortisol.
Treatment may include radiation or surgery. Steroid-induced Cushing's disease can be reversed following withdrawal of the steroid drugs.
Addison's disease results from reduction in glucocorticoid secretion, which may be caused by infection or autoimmune disease. In the latter, proliferation of immunocytes against specific antibodies occurs because of a deficiency of immune suppresser cells in the adrenocortical cells. This causes chronic reduction in the levels of aldosterone and cortisol. Reduced aldosterone secretion decreases renal sodium and water reabsorption, leading to hypovolaemia and

hypotension. In turn this leads to hyperkalaemia, since potassium is retained by the kidney. Autoimmune destruction of melanocytes causes vitiligo.
Treatment of Addison's disease involves replacement of the missing glucocorticoid and mineralocorticoid, usually using cortisol or prednisolone plus fludrocortisone.

Part 2

Q8 Brian's symptoms are not consistent with a diagnosis of either diabetes mellitus or hypothyroidism. He has no glycosuria and, although his blood sugar is high, it is only slightly above normal. The symptoms which suggested diabetes may be due to increased plasma glucose concentration since steroids, such as cortisol, inhibit the actions of insulin. His plasma insulin concentration is normal.
Although Brian is tired and weak, with a puffy face and significantly increased body weight, all of which may be observed in myxoedema (a severe form of adult hypothyroidism), his thyroid hormones and TSH concentration are normal. In hypothyroidism T_3 and T_4 concentrations are usually low, which leads to an increase in TSH because of reduced negative feedback between the thyroid hormones and the anterior pituitary gland.

Q9 There is a marked variation in cortisol secretion over a 24-h period. It is highest in the morning and lowest at night. One single urine or blood sample taken at some point in the day is not likely to allow an accurate estimate of the daily amount of cortisol secreted by the patient to be made.

Q10 Cortisol secretion is normally suppressed by exogenous glucocortocoids. In a normal subject, administration of the synthetic steroid dexamethasone produces a rapid feedback inhibition of CRF and ACTH, which in turn suppresses or diminishes cortisol production to a low level. In Cushing's disease little or no suppression will occur.

Q11 Aldosterone is concerned with regulation of body fluid volume and sodium content and is controlled by two main mechanisms. When extracellular fluid (ECF) or blood volume decreases, or when sodium concentration in the ECF or blood pressure decreases, the enzyme renin is released from the juxtaglomerular tissue of the kidney. Renin converts angiotensinogen to angiotensin I, which is, in turn, converted to the potent vasoconstrictor agent angiotensin II by angiotensin converting enzyme (ACE). Angiotensin II stimulates aldosterone release from the adrenal cortex.
Aldosterone secretion is also stimulated by increased plasma potassium concentration. Potassium is secreted into the urine in exchange for reabsorption of sodium in the distal nephron. Aldosterone also promotes secretion of hydrogen ions from the distal tubule according to the acid–base status of the

body. The net effect of increased aldosterone secretion is increased reabsorption of sodium and loss of potassium in the urine, leading to hypertension and hypokalaemia.

Q12 Cushing's disease is a serious condition which alters the ability of the body to respond to stress and infection and results in tiredness and weakness with increased susceptibility to infections. It is not likely that a patient with these problems would be working optimally or even normally.
Brian has suffered changes in his mental status which have altered his relationship and communication with customers, and employees. These changes can include irritability, depression and paranoia, a feeling that people are always 'getting at him'. Brian's appearance and friendly, helpful attitude is likely to have deteriorated during his illness. So his poor sales figures and failing business can certainly be attributed to the effects of the medical condition described in this study.

Key Points

- The adrenal cortex secretes steroid hormones, primarily glucocorticoids and mineralocorticoids. Deficiency of these steroids results in Addison's disease, characterized by hypoglycaemia, lethargy and weight loss, anorexia and nausea and hypotension, which can be severe and may result in cardiovascular collapse. Treatment includes replacement of the missing steroids.
- Over-secretion of the glucocorticoids, particularly cortisol, causes Cushing's disease, with changes in distribution of body fat, muscle wasting and weakness, osteoporosis, immune suppression with frequent infections, poor wound healing and mood changes, which include depression and sometimes paranoia.
- Cushing's disease is caused by an increase in circulating ACTH, usually from a tumour either in the pituitary gland or at other body sites, or there may be a spontaneous increase in release of cortisol.

CASE STUDY 17 The thirsty schoolboy

Q1 The major hormones involved in blood glucose control are insulin and glucagon.

Q2 Hormones which affect blood glucose concentration are secreted from cells of the islets of Langerhans in the pancreas. Insulin is produced by 60% of the cells (beta cells, or β-cells) and glucagon by 20% of the cells (alpha-cells, or α-cells). Increase in the level of blood glucose initiates the release of insulin. Insulin decreases blood glucose concentration: its main metabolic effects are on the liver, muscle and adipose tissue. The principal organ of glucose homeostasis is the liver as it absorbs and stores glucose as glycogen, releasing it between meals to supply peripheral tissues. Although brain tissue is a major consumer of glucose, usage is not dependent on insulin. Insulin influences the metabolism of glucose in many tissues, particularly muscle and fat cells. It facilitates the transport of glucose, a number of amino acids and phosphate, potassium and calcium across the cell membrane into the cell. The action of glucagon is opposite to that of insulin: it increases the level of glucose in the blood by stimulating gluconeogenesis and glycogenolysis.

Q3 Blood glucose concentration is normally 3.3–5.5 mmol l^{-1} during fasting.

Q4 Diabetes may be regarded as a group of diseases which have glucose intolerance in common. The major types are: type 1 (insulin-dependent) and type 2 (non-insulin-dependent). There is also type 3 diabetes, which affects people in midlife; it has no connection with being overweight. People with type 3 diabetes have an active lifestyle, do not respond to diet control and maintain the normal level of glucose by insulin. Its underlying pathophysiology is not well understood. Gestational diabetes occurs when glucose intolerance develops during pregnancy; it usually returns to normal after delivery. Gestational diabetes can have serious consequences for mother and baby if left untreated. It occurs in 2–5% of all pregnancies, but the incidence is higher in some ethnic groups.

Q5 Approximately 10% of diabetic patients have type 1 diabetes, which usually occurs in childhood with a peak incidence between 10 and 13 years of age. In such patients, up to 80–90% of the β-cells have been damaged, leading to insufficient production of insulin. Insulin is essentially an anabolic hormone, favouring glycogenesis, synthesis of proteins in cells and synthesis and storage of lipids in adipose tissue. As a result of insulin deficiency, glucose accumulates in blood and appears in the urine (glucosuria) because the renal threshold for glucose is exceeded. The resulting symptom is polyuria. Polydipsia (intense thirst) occurs because loss of glucose in urine removes water from the body osmotically, causing thirst and dehydration. An insufficient amount of insulin results in cessation of protein synthesis, with subsequent wasting of muscle cells. As a consequence, large amounts of amino acids are released into the plasma. Polyphagia (excessive hunger), with weight loss and fatigue, occurs as the synthesis of protein in cells decreases and catabolic processes reduce lean body mass and lipid stores. The movement into and storage of lipids in adipose tissue is diminished and lipolysis increases, and so plasma fatty acids increase. The result of enhanced delivery of lipids to the liver is an increase in gluconeogenesis.

Q6 When blood glucose falls, glucagon is released. Glucagon increases glucose output by the liver and stimulates hepatic conversion of amino acids to glucose. To make glucose available for cellular activities, the body produces gluconeogenic enzymes and promotes glycogenolysis and new glucose formation (gluconeogenesis).

Q7 Ketoacidosis is a serious complication of diabetes mellitus. Because of insulin deficiency and consequent increased availability of fatty acids to the liver, the liver overproduces alpha-hydroxybutyrate and acetoacetic acid, increasing ketone production. The ketones are released into the circulation. They are strongly acidic and, when not effectively buffered, cause metabolic acidosis. Coma may then occur because of severe depression of the nervous system.

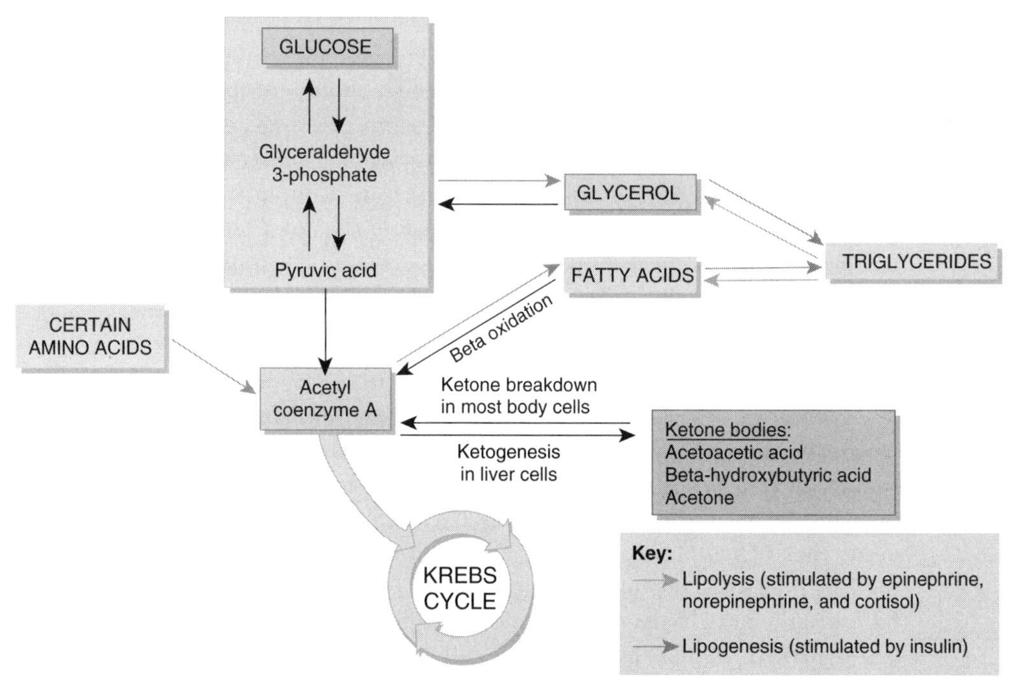

From Tortora and Derrickson *Principles of Anatomy and Physiology, Eleventh Edition 2006.* Reproduced with permission of John Wiley & Sons, Inc.

Except for acetone, all ketone bodies carry a negative charge and are organic acids which are excreted from the body in urine. Owing to their negative charge, ketone bodies remove sodium and potassium ions from the body as they are excreted. Loss of these cations results in electrolyte imbalance, which is associated with vomiting, abdominal pain and dehydration. Patients' breath is likely to smell fruity since acetone, which is more volatile than other compounds, is excreted via the lungs.

Q8 In type 1 diabetes, because of a lack of insulin, a high level of triglyceride is stored in the liver and can subsequently be converted to phospholipids and cholesterol. Hepatocytes synthesize VLDLs, which can be converted to other types of lipoproteins. These lipoproteins are major sources of cholesterol and triglycerides for most other tissues. They leave the liver, enter the blood and can result in rapid development of vascular atherosclerosis. Increased levels of atherogenic oxidized low-density lipoproteins (LDLs) are seen in hyperglycaemic individuals and contribute to macrovascular disease, which is a complication of diabetes mellitus.

Q9 The child's symptoms suggest that he suffers from type 1 diabetes mellitus.

Q10 In symptomatic patients a single elevated blood glucose concentration of $>11.1\ \mathrm{mmol\,l^{-1}}$ glucose is indicative of diabetes mellitus. A confirmatory glucose tolerance test can be used if necessary. After an overnight fast of approximately nine hours, a sample of blood is taken for estimation of fasting blood glucose. The patient then drinks a small volume of fluid containing 100 g (or 75 g) glucose and further blood samples are taken at 30-minute intervals for two hours. A diagnosis of diabetes is made if glucose in the fasting sample is $>6.7\ \mathrm{mmol\,l^{-1}}$ and/or the level in the later sample is $10-11.1\ \mathrm{mmol\,l^{-1}}$ or more. In diabetes mellitus, blood glucose rises to a higher level than normal in this test. In healthy people blood glucose falls rapidly to resting levels in two hours. But in diabetic patients it takes much longer to return to fasting levels (three hours or more).

Q11 There is both a genetic and environmental component in type 1 diabetes. The pathological basis of the condition is autoimmune destruction of the pancreatic islet cells, which is said to be associated with genetic and environmental factors such as viral infection. It has been shown that antibodies to islet cells and insulin autoantibody (IAA) can exist for years before the occurrence of symptoms, possibly as a result of the autoimmune processes: the IAA may form during the process of active islet and β-cell destruction. Both insulin and glucagon play a role in the development of hyperglycaemia and hyperketonaemia, since both α- and β-cell functions are abnormal in diabetes. Both a lack of insulin and a relative excess of glucagon coexist in type 1 diabetes, and so the metabolic abnormalities that occur are likely to be caused by both hormones.

Q12 Insulin injections, which are intended for subcutaneous administration, are prescribed for patients suffering from type 1 diabetes, since gastric acid and enzymes destroy insulin if it is given orally. More than 30 different insulin products are available: rapid-acting, intermediate, long-acting and biphasic; so treatment can be tailored to the requirements of individual patients. The only difference between the products is their time of onset, peak serum levels and the duration of action, which can be 3–24 hours, depending on the preparation.

Q13 The overall aim of treating patients with insulin is to maintain optimal blood glucose control and avoid hypoglycaemia or hyperglycaemia; good control greatly reduces the risk of diabetic complications. This is best achieved via blood glucose monitoring by the patient, which can rapidly detect developing hypo- and hyperglycaemia. In addition patients benefit from education in healthy nutrition, weight control and physical activity. Patients taking insulin may develop hypoglycaemia and some people lack significant warning signs. When blood glucose decreases to $<3.0\,\text{mmol l}^{-1}$ patients usually experience symptoms of sweating, tremor and pounding heart, some will go pale and drowsy and others can quickly fall into a comatose state. Hypoglycaemia can be caused by insulin, missing a meal, the effects of other drugs or infection. Patients taking insulin should avoid alcohol, non-selective β-blockers, oral hypoglycaemics, monoamine oxidase inhibitors (MAOIs) and salicylates, since they can promote the development of hypoglycaemia. Drugs such as thiazides, thyroid hormones, oral contraceptives, corticosteroids, lithium, diazoxide and loop diuretics are associated with hyperglycaemia.

Q14 The presence of hypoglycaemia should be considered in any unconscious patient. If an insulin overdose is suspected, 10–20 g glucose is given by mouth to conscious patients, for example as sugar lumps. In unconscious patients a 50 ml bolus of 50% glucose solution can be administered intravenously, or 1 mg glucagon administered subcutaneously, intramuscularly or intravenously.

Q15 As a rough guide to successful blood glucose control, urine dipsticks can be used (e.g. Clinistix or Clinitest). If a patient has consistent negative tests for glucose and ketones in the urine and reports no symptoms of hypoglycaemia, he or she is probably reasonably well controlled. However, a better test, which assesses long-term control of blood glucose, is estimation of glycosylated haemoglobin (Hb) in blood. The glycosylated Hb measured is expressed as a percentage of normal Hb. Glycosylation involves the formation of a covalent bond between glucose and the beta-chain of Hb; the rate of this reaction is related to the prevailing glucose concentration.

In a healthy individual only 4–8% of the Hb is glycosylated; if blood glucose has been inappropriately high for a period, this percentage will increase. The extent of Hb glycosylation is therefore related to the blood glucose concentration to which a red cell has been exposed over its lifetime (approximatelyx 120 days).

Key Points

- Diabetes may be regarded as a group of diseases which have glucose intolerance in common.
- The major types are: type 1 (insulin-dependent), and type 2 (non-insulin-dependent). There is also a type 3 diabetes, which affects people in midlife and has no connection with being overweight.
- In type 1 diabetes the β-cells have been damaged. The resulting symptoms are polyuria, polydipsia (intense thirst), polyphagia (excessive hunger), with weight loss and fatigue, gluconeogenesis, ketoacidosis and vascular atherosclerosis.
- Both insulin and glucagon play a role in the development of hyperglycaemia and hyperketonaemia, since both α- and β-cell functions are abnormal in diabetes.
- Both a lack of insulin and a relative excess of glucagon coexist in type 1 diabetes and, therefore, the metabolic abnormalities that occur are likely to involve both hormones.
- Pharmacological treatment involves insulin injections.

CASE STUDY 18 Eric's expanding waistline

Q1 It is likely that Eric has type 2 diabetes (non-insulin-dependent diabetes), which is much more common than type 1. The incidence is rising in Western populations as obesity increases.

Eric is overweight, with a body mass index (BMI) of 44 (normal range 18.4–25 in males) and central obesity; he takes little exercise and suffers fatigue and repeated skin infection. All these points are consistent with increased risk of type 2 diabetes. Possibly his nocturia is associated with hyperglycaemia and glucosuria, but it could simply be a result of the large volume of beer taken during the evening. Thirst could also be associated with hyperglycaemia, but may be associated with his 'fondness' for alcohol. His appointment for a glucose tolerance test suggests that his family doctor is considering a diagnosis of type 2 diabetes.

Q2 In normal healthy individuals blood glucose levels are closely regulated and rarely fall outside the range 3.5–8.0 $\mathrm{mmol\,l^{-1}}$, even when fasting or after eating a large amount of sugary food. Diabetes can usually be diagnosed on the basis of a random blood glucose test. But where there are few conclusive symptoms of the condition, a glucose tolerance test may be performed to confirm the diagnosis. In the glucose tolerance test, patients whose fasting blood glucose level is $>6.7\,\mathrm{mmol\,l^{-1}}$ and/or whose two-hour value is $>11.1\,\mathrm{mmol\,l^{-1}}$ are diagnosed as having diabetes. Following the glucose load, diabetic patients show much higher blood glucose values than healthy individuals. While the time taken for the glucose level to fall is less than two hours in a healthy person, it may take more than three hours in a diabetic patient.

Q3 Patients are usually over 40 years of age at diagnosis, and the majority are overweight or obese. Unfortunately, young, obese people are now developing this type of diabetes. Approximately 90% of diabetic individuals suffer from type 2, non-insulin-dependent diabetes. Patients with type 2 diabetes show some classic hyperglycaemic symptoms, with similar consequences to type 1 diabetes. However, if ketone bodies are present, they are only in very low concentration in blood and urine and this very rarely results in coma.

Obesity, particularly central obesity, seems to be a major factor for the development of type 2 diabetes, particularly in genetically susceptible individuals. The condition is more common in southern Asians, people of African and Caribbean ancestry and American Indian populations than in Caucasian people. Type 1 diabetes on the other hand is insulin-dependent and is initially diagnosed mainly in young people. Hyperglycaemia, glucosuria and thirst are prominent with resulting polyuria and polydipsia. Significant weight loss and dehydration occur and the individual may develop ketoacidosis and ketonuria because of the increased release of fatty acids with associated ketone production.

Q4 In type 2 diabetes there is a reduction in the responsiveness of beta-cells (β-cells) to plasma glucose levels (which might be due to a reduction in the number of β-cells or their abnormal function) and an increase in the secretion of glucagon. Many patients with this condition show resistance insulin. *Insulin resistance* is defined as a suboptimal response to insulin in insulin-sensitive tissues (liver, muscle and adipose tissues). Insulin resistance is increased by obesity, inactivity and age. Obesity and insulin resistance coexist in approximately 60% to 80% of patients with type 2 diabetes in the West. In approximately 10% to 40% of patients with type 2 diabetes, amyloid deposits have been found in the islet tissues of the pancreas. Interestingly, the presence of amyloid correlates positively with the age of the patient and the duration and severity of the disease.

Q5 Diabetic states, particularly hyperglycaemia and hyperlipidaemia, are associated with increased atherosclerosis and other deleterious changes in the arteries. There are many contributing factors, including: glycosylation of Hb, which contributes to vessel and tissue hypoxia, decreased high-density lipoproteins, increased LDLs and oxidized LDL with altered vascular endothelial cell function, which promote deposition of atherosclerotic plaques and proliferation of vascular smooth muscle cells. The latter leads to increased vascular pressure and to macrovascular disease, which is a major cause of morbidity and mortality. The incidence of coronary artery disease (CAD) is higher in diabetic people than in the general population, even when other common risk factors, such as hypertension and hyperlipidaemia, are taken into consideration. Stroke is twice as common in those with diabetes than in the general population. There is also an increased incidence of peripheral vascular disease in diabetes. Diabetic patients have a greater risk of gangrene and amputation than the non-diabetic population.

Q6 Microvascular complications in diabetes are a major cause of end-stage renal failure. Among the changes to small blood vessels in diabetes are thickening of the basement membrane and proliferation of endothelial cells. These changes, which appear to be associated with hyperglycaemia and glycation of structural proteins, eventually lead to decreased tissue perfusion, hypoxia and ischaemia. In the kidney the glomeruli as well as the blood vessels are damaged. The early changes to the kidney are usually without symptoms and begin to develop after 5–10 years of the diabetic state. The first sign of renal dysfunction is usually microalbuminuria, a very small loss of albumen in the urine; later, after about 25 years, glomerular sclerosis, reduced glomerular filtration, hypertension and uraemia develop. When patients strictly control their blood sugar levels, these renal complications are greatly reduced.

Q7 Diabetic neuropathy is thought to result from multiple genetic, metabolic and environmental factors. In the early stages there is slowed motor and sensory nerve conduction. Later, axonal degeneration may occur: sensory nerves appear to be affected early leading to pain, alteration or loss of sensation, for example tingling and numbness in the feet. These changes in sensation in the feet may affect balance. Diabetic patients often develop foot problems, related to the combined effects of ischaemia, neuropathy and infection. In severe cases gangrene and amputation may follow.

The ANS is also affected and postural hypotension may occur since sympathetic control of blood vessels is lost. Other areas affected by neuropathy include the gut, urinary bladder and sexual function, leading to diarrhoea, incomplete bladder emptying and impotence respectively.

The retina is the most metabolically active tissue in the body and so is very vulnerable to the microvascular changes which occur in diabetes. Diabetes affects the eyes in a number of ways; the most common is diabetic retinopathy, which involves increased thickness of the retinal basement membrane and increased permeability of its blood vessels. The severity of the retinopathy is related to the age of the patient, duration of the diabetic state and extent of glycaemic control. Later changes in the eye include macular oedema and retinal ischaemia, which threaten the sight of the patient. All these deleterious changes are minimized if blood glucose is tightly controlled.

Q8 The major focus of management is to prevent long-term complications and so monitoring blood glucose is essential. Non-pharmacological management includes: weight reduction, exercise, appropriate diet, correction of hyperlipidaemia and hypertension, and avoidance of smoking, which are all recommended. A sensible diet can reduce the fasting blood glucose level to $<6\,\text{mmol l}^{-1}$ in about 15% of patients. Patient education and support are integral to success.

Q9 Oral hypoglycaemic agents can help a further 50% of patients to reduce their blood glucose. However, in order to be effective, they require some remaining function in the pancreas. These drugs have their greatest effect on basal blood glucose concentration and have little effect on the raised blood glucose that occurs after food intake; thus, they should be used in combination with a sensible diet. The following drugs are commonly used: biguanide drugs, such as metformin, and sulfonylurea drugs, such as chlorpropamide, glibenclamide, tolbutamide and glipizide.

Q10 Sulfonylurea drugs enhance insulin secretion. They bind to the receptors on the β-cells of the islets of Langerhans, causing a partial depolarization of the cell membrane, influx of calcium ions and a reduction in potassium efflux. The ultimate result is increased insulin secretion. These drugs also increase the sensitivity of the β-cells to stimuli that cause insulin secretion, possibly by increasing the intracellular levels of a cyclic nucleotide second messenger following the inhibition of phosphodiesterase. In high concentrations, sulfonylureas also increase insulin-mediated tissue uptake of glucose and increase insulin receptor density in tissues, although these actions are probably irrelevant at therapeutic doses. Dosage depends on the individual drug, for example for glibenclamide the starting dosage is 5 mg per day.

Biguanide drugs, such as metformin, act both by increasing the peripheral uptake and utilization of glucose and by reducing gluconeogenesis. As with the sulfonylureas, some functioning of the islet cells and the presence of insulin is necessary for this beneficial effect. The dosage of metformin is 500 mg per day for one week, then 500 mg twice daily for one week, building up to 2 g daily in divided doses.

Q11 There is some risk of hypoglycaemia with the long-acting sulfonylureas, and weight gain is also associated with the use of these drugs. They are used only if diet alone has failed to control blood glucose and symptoms. Metformin is usually prescribed for overweight and obese patients. The use of the biguanide drug metformin is associated with gastrointestinal side effects at high doses; it may possibly cause profound lactic acidaemia in some patients, particularly in those with renal impairment.

Q12 Thiazolidinediones, such as pioglitazone and rosiglitazone, reduce tissue resistance to insulin and can be considered to be insulin sensitizers. Use of these agents reduces blood sugar concentration and they are particularly useful when sulfonylureas or metformin no longer produce an adequate response because of diminishing insulin release. There have been a few reports of liver toxicity associated with the use of thiazolidinediones, so their use in patients with impaired liver function is contraindicated. However, these agents have been introduced relatively recently and their long-term effects have not yet been adequately demonstrated.

Key Points

- Type 2 diabetes is associated with obesity and is most common in people over 40 years of age. The pancreatic β-cells show a reduced response to blood glucose levels, but patients do not normally require insulin injections.
- Patients present with hyperlipidaemia and hyperglycaemia with impaired glucose tolerance, glucosuria, polyuria and polydipsia.
- Ketosis rarely occurs.
- The long-term consequences of type 2 diabetes are similar to those of type 1 diabetes. These include: macro- and microvascular damage, leading to

renal dysfunction and failure, ocular damage, leading to visual impairment, and diabetic neuropathy involving mainly the ANS.

- Close control of blood glucose greatly decreases diabetic complications and the recommended management of this condition involves increased exercise with appropriate diet. Treatment may also include hypoglycaemic agents, such as sulfonylureas or metformin, or glucose sensitizers, such as thiazolidinediones, when diet and exercise are insufficient to control hyperglycaemia.

4

Cardiovascular disorders

CASE STUDY 19 Annie's heartache

Part 1

Q1 Coronary arteries are the first to branch off the aorta. The heart has a large blood flow (200 ml min^{-1}) but also has great metabolic needs and so has a relatively poor oxygen supply, with little in reserve when oxygen demands increase. At each heart beat (systole), the coronary arteries are compressed by cardiac contraction and blood flow diminishes to a low level. This effect is very marked in the left ventricle, and over 80% of coronary flow to the left ventricle occurs in the periods between beats (diastole). When heart rate increases (tachycardia), the duration of diastole decreases much more than the duration of systole, and the period available for perfusion of cardiac muscle diminishes.

Coronary blood flows through the cardiac muscle in arteries, capillaries and veins and returns to the heart via the coronary sinus. Extraction of oxygen from coronary blood is high and, when oxygen requirements increase, blood flow must also rapidly increase to supply the extra oxygen needed.

Q2 The heart is supplied by both divisions of the autonomic nervous system (ANS) via the vagus and sympathetic nerves. The vagus innervates the sinoatrial (SA) and atrioventricular (AV) nodes and atria, but very few fibres

Clinical Physiology and Pharmacology Farideh Javid and Janice McCurrie

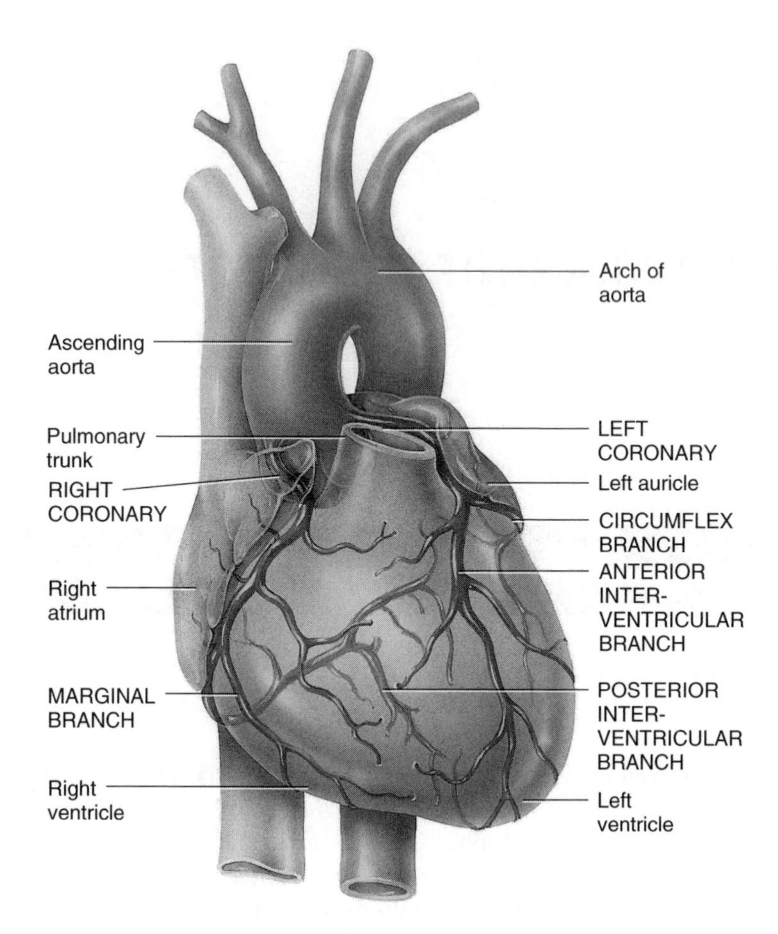

Anterior view of coronary arteries. From Tortora and Derrickson *Principles of Anatomy and Physiology, Eleventh Edition 2006*. Reproduced with permission of John Wiley & Sons, Inc.

innervate the ventricles, so vagal stimulation has little impact on ventricular contraction. Sympathetic fibres innervate the nodes, atria and ventricular muscle. Coronary blood vessels have a rich sympathetic innervation, but stimulation of these nerves has little direct effect on the control of coronary blood flow, as the effects of autoregulatory mechanisms are much more powerful. Coronary vessels are also innervated by nitrergic and peptidergic nerves.

Q3 The major control in the coronary circuit is autoregulation, predominantly hypoxia. A small degree of hypoxia dilates coronary vessels and produces a large increase in blood flow: increased arterial PCO_2 and decreased pH also dilate coronary arteries and increase blood flow.

Q4 Angina pectoris (pain in the chest) occurs when the oxygen available to cardiac muscle is inadequate for its needs, that is when there is myocardial ischaemia. There may be a chemical factor, released during ischaemia, which initiates a sensation of pain. The pain is substernal but may radiate down the left arm and up to the chin and teeth on the left side.

Typically, angina occurs when patients who have narrowed coronary vessels start to exercise. Exercise increases heart rate and cardiac output, increasing the myocardial requirement for oxygen. Narrowed coronary arteries may not be able to dilate sufficiently to permit the increased blood flow and oxygen delivery required by cardiac muscle during exercise. The most common reason for narrowed coronary blood vessels is the build-up of fatty deposits (atheroma) on the inner wall of coronary vessels, which reduces the space available for blood flow.

Typically, angina presents as pain, but sometimes patients experience a sensation of chest tightness or heaviness. These symptoms are made worse by heavy meals, emotional stress and by cold weather, since the latter causes peripheral vasoconstriction and increases the workload of the heart.

Q5 Angina is usually immediately preceded by changes in the ST segment of the electrocardiogram (ECG). The typical ECG change is a depression of the ST segment, but in severe angina caused by coronary artery spasm the ST segment may be elevated for a short time. Twenty-four-hour ECG monitoring has shown that some patients have 'silent' myocardial ischaemia, episodes of ST depression which occur without appreciable chest pain or tightness.

Q6 GTN (glyceryl trinitrate) is an organic nitrate which causes smooth muscle relaxation, particularly of venous muscle. The value of venous dilation is that it reduces cardiac preload, which decreases cardiac energy expenditure, so reducing oxygen requirements. When GTN enters the smooth muscle cell, it is metabolized to release nitric oxide (NO), a powerful natural vasodilator; this process requires the presence of free -SH (sulfhydryl) groups. Repeated or prolonged administration of nitrate drugs can result in development of tolerance with reduced vascular relaxation, which may be due to reduced availability of the cellular -SH groups. The NO released in vascular muscle activates guanylate cyclase, by combining with the haem group, and increases cyclic guanylate monophosphate (cGMP) concentration. Protein kinase G, dephosphorylation of the myosin light chain, ion channel activity and other proteins are affected by the increased cGMP, leading to relaxation of the muscle cell.

Q7 The major adverse effects of the nitrates are related to their relaxant action on venous muscle. Patients may experience enhanced venous pooling when standing up from a supine or sitting position, which causes postural hypotension and dizziness. Flushing of the face and headache may also be experienced. In very rare cases the nitrate oxidizes significant amounts of the ferrous iron in haemoglobin to the ferric form, producing methaemoglobin, which does not function as an oxygen carrier.

Part 2

Q8 The most common type of angina is *stable angina*, which is induced by exercise, cold, heavy meals and emotional stress. This is well controlled by GTN and other anti-anginal agents. Two other types exist: unstable angina and Prinzmetal, or variant, angina. *Unstable angina* refers to angina which is increasing in severity and frequency. It may start at rest or after minimal exercise or stress and is associated with thrombus formation on ruptured atheromatous plaques, which partly occlude the coronary vessels. Patients with this form of angina are at substantial risk of developing full occlusion of a coronary artery and myocardial infarction. *Prinzmetal angina* is a less common condition and is generally associated with spasm of the coronary arteries. It may occur in patients with minimal atherosclerotic damage and rarely progresses to infarction.

Q9 Following absorption from the gut, GTN is extensively metabolized by the liver, which greatly limits its availability. GTN must be administered as a buccal spray or sublingual tablet, or as a skin patch.

Q10 A beta-adrenoceptor (β-adrenoceptor) antagonist would *not* be suitable for treatment of a patient, such as Annie, who suffers from asthma, as it can antagonize the relaxant effects of β_2-receptors as well as the stimulant β_1-adrenoceptor on cardiac muscle. Even the selective β_1-adrenoceptor antagonists have some action on bronchial muscle, which may result in bronchoconstriction.

Q11 Propranolol is a non-selective β-adrenoceptor antagonist: it blocks effects of sympathetic stimulation at both β_1- and β_2-receptors. Antagonism of β_2 effects causes vasoconstriction and bronchoconstriction. When given to a patient at rest, propranolol causes only a small reduction in heart rate and cardiac output and little immediate change in blood pressure (BP). But it is beneficial to some patients with stable angina since it reduces the response of the β_1-adrenoceptors in the heart to sympathetic stimulation during exercise and stress, so reducing the work of the heart and the incidence and severity of angina.

Q12 Stable angina usually subsides over several minutes when exercise is stopped and the patient's usual medication, such as GTN, is taken. Pain which does not diminish, accompanied by pallor, hypotension, shock, nausea and sweating, suggests there may be a coronary occlusion.

The diagnosis of myocardial infarction is based on the patient's history, ECG changes and significant increases in myocardial enzymes. In particular there is a rise in cardiac creatine phosphokinase, reaching a peak 24 hours post infarction and an increase in troponin I and T, which are specific markers of myocardial injury when infarction has occurred.

Q13 Calcium channel blockers decrease the opening of L-type calcium channels in the plasma membrane of vascular smooth muscle cells, and so reduce intracellular calcium concentration and contractile activity. The blood vessels therefore dilate. Calcium channel blockers act mainly on the arterial side of the circulation, and the dihydropyridines, such as nifedipine, are useful coronary arteriolar dilator agents. These agents are usually the treatment of choice for Prinzmetal angina.

Key Points

- Cardiac muscle has a large oxygen requirement, and extraction of oxygen from coronary blood is high. The coronary arteries are compressed in systole, particularly in the left ventricle. Sympathetic stimulation, which increases heart rate and force, reduces the duration of diastole and increases myocardial oxygen consumption. A slow heart rate improves coronary perfusion, reduces oxygen demand and is beneficial for coronary perfusion.
- The major control on the coronary circulation is autoregulation.
- Anginal pain occurs when the oxygen supply to the myocardium is inadequate, and is most commonly experienced during exercise. Some patients experience angina at rest; this appears to be associated with spasm of the coronary arterial muscle.
- Not all patients with angina experience pain; some patients report only chest tightness or ache.
- The most common medication for angina is GTN; this is available as buccal tablets, sprays or skin patches. Calcium channel blockers and beta-blockers are also effective anti-anginal agents.

CASE STUDY 20 The executive's medical check-up

Q1 The normal range of BP for adults over the age of 18 years is 120–140 mmHg systolic and 75–80 mmHg diastolic. However, BP increases with age and body weight, varies between individuals and in a single individual BP naturally varies during a 24-hour period. In new guidelines for management of high BP (hypertension), the target pressure to be achieved in treating hypertensive patients has been revised downwards because there is clear evidence that high BP is associated with increased risk of heart failure, vascular and kidney disease and stroke.

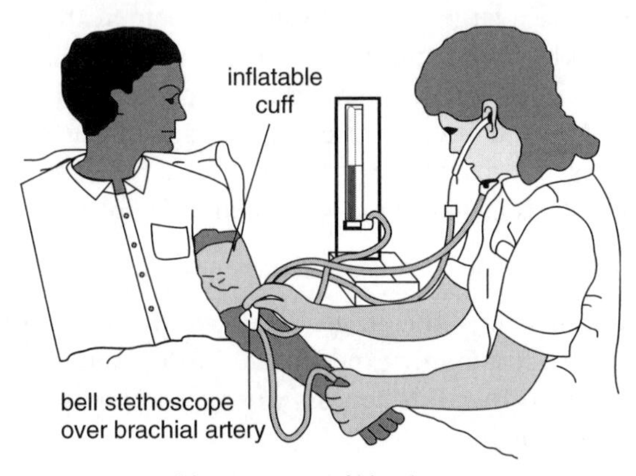

Measurement of blood pressure

Sam's BP two years ago was higher than normal, but could have been lowered by lifestyle changes such as weigh reduction, increased exercise and salt restriction. Now he has gained more weight and his BP is likely to be higher than before.

The short-term mechanism controlling BP from minute to minute involves arterial baroreceptors. When changing body position, baroreceptors detect changes in BP and elicit reflex responses via the cardiovascular centre in the medulla, which reverse the change and return BP to the original level. Baroreceptors operate these reflexes in hypertension, but adapt to the increased pressure so that they operate around a higher set point.

The long-term control of BP over weeks or months involves the renin–angiotensin system and the kidney, which regulate the salt and water content of the body. Blood volume is a major determinant of BP, and if blood volume increases because of salt and water retention BP also increases. Similarly, salt and water depletion or blood loss causes decrease in both blood volume and pressure.

Q2 The heart and blood vessels are innervated by sympathetic nerves. The 'fight or flight' response of the sympathetic system is activated in stress, and epinephrine (adrenaline) is released into the blood. These responses increase the rate and force of the heart and constrict many blood vessels, both of which raise BP. In addition, continued sympathetic stimulation eventually causes structural changes in blood vessels, activation of the renin–angiotensin system and procoagulant effects. If stress is prolonged, these responses contribute to a sustained increase in BP.
Sam could decrease the stress in his job by reducing the number of overseas trips he makes and working shorter hours He could stop rushing, slow down a little and take things more calmly.

Q3 Sustained hypertension damages the walls of blood vessels, leading to dysfunction in the tissues perfused. The heart is often affected: ventricular hypertrophy, coronary disease, angina, congestive or sudden heart failure may occur. The brain and eyes can be damaged as exudates and haemorrhages occur in the retina and there may be kidney damage, leading to renal failure.

Q4 Beta-adrenoceptor (β-adrenoreceptors) antagonists were originally introduced as anti-anginal agents. It was then noted that patients' BP decreased over a period of weeks: these agents have now been used to treat hypertension for many years. The beta-blockers (β-blockers) are the agents of choice for young hypertensive patients. Their mechanism of action is unclear, but there are several components:

(1) Reduction in sympathetic stimulation of the heart via antagonism at β_1-adrenoceptors in cardiac muscle.

(2) Antagonism of β_1-receptors on the juxtaglomerular cells of the kidney that reduce the release of renin.

(3) Reduction of sympathetic activity via a central effect in the brain.

Part 2

Q5 BP varies considerably according to the activities and anxieties of the patient and time of day. If a patient rushes in late to a medical appointment and is very anxious or worried, their BP is likely to be much higher than usual. Waiting a few minutes to allow time for the patient to calm down facilitates a more accurate measurement to be made. It is usual to measure BP on more than one visit, to ensure that it is consistently raised, before starting antihypertensive therapy. In Sam's case he was late and rushed to his appointment, which would increase the activity of his sympathetic nervous system and temporarily raise his cardiac output and BP.

Q6 The renin–angiotensin system is involved in long-term regulation of body fluid volume and BP. When the β_1-receptor on juxtaglomerular cells is stimulated by reduced BP or sodium depletion, the proteolytic enzyme renin is released. Renin converts angiotensinogen into angiotensin I, which is then converted to angiotensin II, a potent vasoconstrictor peptide, by angiotensin converting enzyme (ACE). Angiotensin causes release of the salt-retaining steroid aldosterone from the adrenal cortex. Sodium and water are retained by the kidney to increase blood volume and BP.

Q7 Sam's BP has now increased significantly. British Hypertension Society guidelines recommend that patients with systolic pressure >160 mmHg or diastolic pressure >100 mmHg should be treated; so an appropriate antihypertensive agent is now required. Although young patients are treated initially with a β-adrenoceptor antagonist, older patients like Sam are normally started on a thiazide diuretic, unless there is a particular reason or concomitant condition which contraindicates these agents. In Sam's case there is evidence of an increased plasma cholesterol concentration. Thiazide diuretics are safe and effective antihypertensive agents, but tend to raise blood lipid levels. So, for Sam, an agent which reduces the activity of the renin–angiotensin system by inhibiting the production of angiotensin II from angiotensin I, with no adverse effect on blood lipids, appears to be a reasonable alternative. Note: even if Sam had been a much younger person, a β-blocker would not have been suitable for him since he has occasional asthma. Beta-blockers can cause bronchoconstriction and so are contraindicated in asthmatic patients.

Sam did not do well on captopril as he developed an irritating, dry cough. ACE inhibitors not only inhibit conversion of angiotensin I to angiotensin II but also inactivate the degradation of vasodilator bradykinin and other peptides, which may contribute to their effectiveness in lowering BP. Kinin accumulation around the larynx may cause the dry cough that is a common side effect of some ACE inhibitors. These agents also sometimes cause taste disturbances. A more suitable alternative, an alpha-adrenoceptor (α-adrenoceptor) antagonist, which improves the patient's lipid profile by decreasing the low-density lipoproteins and increasing the high-density lipoproteins, was then selected for Sam.

Q8 Alpha-adrenoceptors are found on vascular smooth muscle and mediate vasoconstriction. Blood vessels also possess β_2-receptors, which mediate vasodilation. Cardiac muscle possesses β_1-receptors: stimulation of the latter increases the rate and force of contraction of the heart.

Prazosin is an α_1-selective adrenoceptor antagonist which causes vasodilation and a fall in BP. Alpha-adrenoceptor antagonists also have a favourable effect on blood lipids and are useful for people with raised cholesterol. However, orthostatic hypotension may occur during treatment because prazosin interferes with the postural reflexes, which are triggered when a patient stands up from a supine or sitting position.

Normally, when a person stands up, gravity shifts blood towards the lower parts of the body. As the baroreceptors pick up this slight drop in BP, they elicit reflex vasoconstriction via α_1-receptors in blood vessels and an increase in heat rate via β_1-receptors in the heart. If α-adrenoceptors are antagonized, less vasoconstriction occurs and the BP remains low: sometimes this is pronounced and causes the patient to become dizzy or faint on standing.

Q9 An alternative to the ACE inhibitors is an angiotensin receptor antagonist such as losartan or valsartan. This class of drug also acts on the renin–angiotensin system and appears to offer all the advantages of ACE inhibitors, without causing the dry cough which Sam found unacceptable.
Calcium channel blocking agents, such as verapamil and nifedipine are also satisfactory antihypertensive agents. These drugs reduce the influx of calcium ions into vascular muscle cells following excitation and so cause vasodilation. They act mainly on arterial vessels in the circulation.

Q10 Yes. The target for BP reduction is $<140/<85$ mmHg, and the audit standard is currently $<150/<90$ mmHg. So Sam's BP is within the recommended range and is now satisfactory.

Q11 If this lunch is typical, Sam's diet contains a large amount of fat and sugar with little fruit or vegetables. He could cut down on the saturated fat, sugar and salt in his diet and replace the fried foods with grilled meat or fish, fresh fruit and vegetables. Although a small amount of alcohol is thought to reduce the risk of heart disease, drinking a large amount of wine plus brandy is excessive and, for overall health, should be reduced. These dietary changes, plus introducing some exercise into his daily routine, would help Sam to reduce his weight and BP. If he smokes, reducing the number of cigarettes each day or cutting out smoking altogether would also reduce his BP. Each of the lifestyle modifications alone would reduce his BP a little; added together, they would be a very useful adjunct to the antihypertensive therapy he now requires.

Key Points

- The ideal range for adult BP is 120–140 mmHg systolic and 75–80 mmHg diastolic; BP varies over the day and from day to day.
- The short-term mechanism for BP control centres on the baroreceptor reflex. The long-term mechanism involves the renin–angiotensin system and body sodium control via aldosterone.
- Sustained hypertension damages tissues (end organ damage), particularly the heart, brain, eye and kidney tissues.
- Beta-adrenoceptor antagonists are effective antihypertensives and operate partly by blocking β_1-receptors on the heart and juxtaglomerular cells of the kidney and partly by central actions which reduce sympathetic activity.

- Thiazide diuretics are effective antihypertensive agents at doses lower than required for diuresis. They decrease body water and sodium and have direct dilator actions on arterial blood vessels
- Alpha-adrenoceptor antagonists are useful antihypertensives; they block the vasoconstrictor effects of α-receptor stimulation. There is sometimes a problem with postural hypotension as reflexes which normally raise BP on standing are interrupted. These agents have a beneficial effect on blood lipids.
- ACE inhibitors are powerful antihypertensives, suitable for patients with an increased renin production. Side effects include cough and taste disturbance; angiotensin receptor antagonists possess the advantages of ACE inhibitors without eliciting cough. Calcium blocking agents are also useful antihypertensive agents.

CASE STUDY 21 A hypertensive emergency

Part 1

Q1 BP increases with age; systolic pressure usually rises faster than diastolic pressure. A normal range of BP for a person in the age group 20–30 years is likely to be 110–120 mmHg systolic and 70–80 mmHg diastolic. Resting heart rate is normally 50–70 beats per minute.

Q2 The cause of 90% of hypertension (essential hypertension) is unknown. Risk factors for essential hypertension include:

- family history (genetic factors)
- obesity, lack of exercise and sedentary habits
- alcohol abuse
- high salt intake
- stressful work and lifestyle.

Essential hypertension is a silent pathological process which progresses at a variable rate in different individuals, damaging tissues of the heart, brain, kidney and eye, but usually produces no symptoms. Since the condition is normally symptomless, there is likely to be a large number of undiagnosed hypertensive patients in the community.
Billie might have developed rapidly worsening essential hypertension or might have hypertension that is secondary to another condition. Her headache could possibly be associated with a recent viral infection, unconnected with her hypertension. Whatever the underlying cause, Billie's BP is very high and requires immediate treatment.

Q3 Billie drinks modestly; she is not exceeding the recommended maximum weekly intake for women. In addition there is some evidence that moderate consumption of wine, particularly red wine, can benefit the heart. Although cigarette smoking contributes to overall cardiovascular risks, it does not appear to be directly associated with hypertension, unless it is very heavy.

Q4 For young people with essential hypertension, either a beta-blocker (β-blocker) or an ACE inhibitor is recommended. For older patients, the medication of choice for hypertension is either a diuretic or calcium channel blocker.

Q5 Beta-blockers can have a number of adverse effects. In fact, all drugs used to treat hypertension have some side effects. Beta-adrenoceptor antagonists are no exception, and Billie did not perceive that taking the β-blocker was helping her. Side effects include cold extremities, hypoglycaemia, bronchoconstriction (making them unsuitable for asthmatic hypertensive patients) and sometimes bad dreams or nightmares. Some patients taking β-blockers appear to be particularly affected by fatigue. Since hypertension is itself without symptoms, the benefits of drug treatment may not be apparent to a patient. Drug compliance can therefore be a problem.

Q6 The conditions associated with secondary hypertension include:

- renal disease, including renal parenchymal disease, for example pyelonephritis and renal failure
- tumours of the adrenal medulla, for example phaeochromocytoma
- tumours of the adrenal cortex, for example in Cushing's syndrome
- vascular diseases, such as stenosis of the aorta or renal artery
- pre-eclampsia (in pregnancy)
- iatrogenic disease (one caused by medication), for example oral contraceptives, corticosteroids.

Q7 When hypertension is discovered, the following tests may be recommended:

(1) blood cell count, erythrocyte sedimentation rate and plasma electrolytes

(2) blood glucose, cholesterol, urea and creatinine

(3) examination of urine (e.g. for glucose and albumin)

(4) chest X-ray (to detect ventricular hypertrophy).

The eyes should be examined for retinal changes. If stenosis of an artery is suspected, further tests, scans and angiography are carried out.

Part 2

Q8 A pregnancy test is necessary because hypertension is a feature of pre-eclampsia, a serious condition which can occur in pregnancy and which threatens the life of both mother and foetus. Also, many antihypertensive drugs are contraindicated in pregnancy. It is necessary to know whether the patient is taking prescribed medicines or is self-medicating, as some drugs, such as monoamine oxidase inhibitors (MAOIs), can interact with dietary components to cause a very rapid rise in BP.

Q9 Sodium nitroprusside acts via the production of NO. It is a powerful vasodilator and a potent, rapidly acting antihypertensive agent. The drug is administered by intravenous infusion but is then converted to thiocyanate in plasma. Thiocyanate toxicity can occur with continued use; consequently, sodium nitroprusside can be used only for short-term treatment.

Q10 When the lumen of the renal artery is reduced by >70%, the kidney becomes ischaemic and the renin–angiotensin system is activated. Renal ischaemia causes a reduction in glomerular function and triggers the release of renin from juxtaglomerular cells. Renin acts on angiotensinogen to produce angiotensin I, which is converted to angiotensin II by ACE. Angiotensin II is a potent vasoconstrictor that increases BP. Angiotensin II also releases aldosterone, which stimulates the kidney to retain more salt and water, and so increases extracellular fluid and blood volume. An increase in blood volume results in increased BP.

Surgical removal of a renal artery obstruction usually reduces BP to an acceptable level and any residual hypertension can be easily managed.

Q11 The antihypertensive drugs which interact with the renin–angiotensin system are ACE inhibitors, angiotensin receptor antagonists and β-blocking drugs (which reduce renin secretion via antagonism at the β_1-receptor on juxtaglomerular cells). This group of agents is less effective in patients who have low renin levels. It explains why agents affecting the renin–angiotensin system are less active than diuretics and calcium channel blocking drugs in lowering BP in elderly people and Afro-Caribbeans, who generally have low plasma renin levels.

Hypertensive patients with normal or high renin levels can benefit from treatment with agents which affect the renin–angiotensin system.

Q12 Labetalol has antagonist effects at both alpha- and beta-adrenoceptors. It acts rapidly and is one of the few agents which is safe to use in pregnancy. It both reduces cardiac output and elicits peripheral vasodilation. These actions reduce peripheral resistance and result in the effective lowering of BP.

Key Points

- Ninety per cent of hypertension is essential hypertension which has no symptoms and no known cause. There is a genetic component, and various lifestyle factors increase BP, such as obesity and sedentary habits, stress and a high salt and alcohol consumption.
- The drug of choice in young hypertensive patients is either a β-blocker or an ACE inhibitor. Beta-blockers have several side effects, including

bronchoconstriction, hyperglycaemia, bad dreams and fatigue, which may be marked in some patients. Patient compliance can therefore be a problem.

- Only 5–10% of hypertension has a known cause; in these cases, it is secondary to a condition such as a tumour of the adrenal medulla, pre-eclampsia during pregnancy, renal disease or renal artery stenosis. Removal of the primary cause, such as stenosis of the renal artery, resolves the hypertension.
- Since pre-eclampsia of pregnancy may be a cause of escalating high BP, it is necessary to carry out a pregnancy test in young female patients. Also some antihypertensive agents are teratogenic and so are unsuitable in pregnant patients. Labetalol, an alpha- and beta-adrenoceptor antagonist, is a drug which is safe for treating hypertensive pregnant patients.

CASE STUDY 22 Harry Mann's bad day

Part 1

Q1 Strenuous exercise, stress and strong emotions, such as anger, activate the sympathetic nervous system, increasing heart rate, BP and the oxygen requirements of cardiac muscle. If coronary vessels are narrowed by atherosclerosis, some areas of the ventricle receive insufficient blood and may become ischaemic. This usually causes substernal pain (angina), often radiating up to the jaw and along the left arm. But pain is not always present; some people feel only an ache or tightness. Aches and pains in the chest can also be caused by indigestion or thoracic muscle damage following heavy lifting or other vigorous activities, such as gardening.

The presence of myocardial ischaemia associated with angina is shown on an ECG trace as a depression of the ST segment, or in some cases by inversion of the T wave. These changes resolve following rest, as blood flow once more becomes adequate for the metabolic needs of the heart muscle. However, if a patient has a heart attack and a coronary vessel is completely blocked, chest pain does not diminish with rest and ECG changes do not resolve. Areas of cardiac muscle beyond the block begin to die after about 20 min of ischaemia; these areas release enzymes which are characteristic of damaged or dying myocardial cells. The cardiac isoenzyme of creatine kinase, creatine kinase MB, and lactic dehydrogenase levels rise and remain elevated for some days. The most specific marker for myocardial damage is the presence in plasma of troponin I and T.

A combination of symptoms, elevated myocardial enzymes and ECG changes can confirm that myocardial infarction has occurred.

Q2 Yes. Luckily, Harry does not seem to have suffered a heart attack, but he has developed mild heart failure, shown by cardiac enlargement and swollen ankles. A normal heart can pump out the blood returning to it via the veins. As the heart begins to fail it is unable to maintain this output and the ventricles enlarge, because of additional blood. Venous pressure rises and disturbs tissue fluid formation as a result of increased hydrostatic pressure in the capillaries. More fluid moves out of the capillaries than can be reabsorbed and this leads to tissue oedema, which is most easily observed in the areas of body particularly affected by gravity: the ankles and feet.

Q3 Tissue fluid is formed by capillaries. At the arterial end of the capillary, hydrostatic pressure filters out fluid, without protein or cells, into the tissues. As the blood reaches the venous end of the capillary, the hydrostatic pressure has fallen and the osmotic pressure that is due to the plasma proteins has increased, since fluid has been lost. The increased osmotic pressure in the blood at the venous end now favours return of fluid back into the capillary. This process ensures a continuous circulation of fluid, containing oxygen and nutrients, through the tissues. Disturbance of hydrostatic pressure, at either the arterial or venous end of the capillary, changes tissue fluid formation. In heart failure the increased venous pressure reduces the net force which returns fluid back to the capillary blood; fluid therefore remains in the tissues, forming oedema.

Q4 Harry's BP is not OK; it is high. Although BP naturally increases with age because of changes in the connective tissues of the arterial system, guidelines suggest that antihypertensive treatment should aim to reduce a patient's BP to 140/85 mmHg or below. A lower target BP is recommended for diabetic patients as they are at greater risk of cardiovascular problems, such as myocardial infarction.

High BP stresses the heart by increasing afterload; the ventricle has to pump blood against increasing pressure. The response of ventricular muscle is enlargement (hypertrophy) and increased oxygen consumption: cardiac efficiency is reduced. High BP is a contributory factor of Harry's heart failure. Hypertension is the significant underlying cause of heart failure in approximately 70% of patients. Prolonged hypertension is also an important risk factor for myocardial infarction and stroke.

Q5 Yes. Thiazide diuretic drugs are one of the treatments of choice for hypertension in elderly patients. Bendroflumethiazide, 2.5 mg daily, is commonly prescribed for hypertension in the United Kingdom. Although the thiazides have been in use for many years, their mechanism of action is not completely understood. They reduce renal reabsorption of sodium and water and so initially decrease blood volume; they also dilate blood vessels and BP falls. However, blood volume may return to normal while the vasodilation and antihypertensive action remains.

Q6 Thiazide diuretics are moderately powerful diuretic agents acting on the distal tubule of the nephron. They reduce reabsorption of sodium chloride and water by blocking the electroneutral sodium chloride (NaCl) transporter system at the luminal border of the distal tubular cells. In addition there are direct relaxant effects on vascular smooth muscle which reduces BP. Diuretics help patients in heart failure by reducing peripheral oedema and decreasing blood volume, which in turn reduces BP. In this way both preload and afterload are decreased and the work of the heart is diminished.

Part 2

Q7 Heart failure is characterized by poor ventricular function and decreased stroke volume. One ventricle only may be affected, but often both ventricles show reduced function. Patients in failure are breathless, have reduced capacity to exercise and suffer from fatigue. Harry has signs of right ventricular failure: he suffers from nausea, loss of appetite and abdominal discomfort with bloating. These symptoms are characteristic of systemic venous congestion and their presence suggests that Harry has now developed congestive heart failure.

Q8 In addition to the intended therapeutic effects, thiazide diuretics can have adverse effects of hypokalaemia, hyperglycaemia and hyperuricaemia. These are not often observed when the usual low dose of thiazide is used. If the dosage is increased, the therapeutic effect is not greatly enhanced, but the likelihood of adverse effects increases considerably. It is therefore better to change to a more powerful agent, such as a loop diuretic, than to increase the dose of the thiazide.

Q9 Loop diuretics, such as torasemide, are known as *high-ceiling diuretics* as they are very powerful drugs, capable of producing a large salt and water loss: 15–25% of filtered NaCl and water may be excreted. Loop diuretics inhibit the mechanism in the ascending limb of the loop of Henle, which pumps Na^+, K^+ and Cl^- ions out of the tubule without accompanying water. This mechanism is central to the counter-current multiplier system that creates and maintains osmotic gradients in the renal medulla, allowing concentration of the urine. Inhibition of ion transport in this part of the tubules reduces the medullary osmotic gradient and delivers large amounts of solute to the distal nephron, allowing excretion of a large volume of salt and water. Furosemide is the most commonly used agent in this group.

Because of its powerful action in the kidney, torasemide can rapidly deplete extracellular fluid volume and mobilize tissue oedema, particularly from the lung. This both improves the patient's breathing and improves their ability to exercise.

Q10 Loop diuretics have similar side effects to the thiazides: hypokalaemia, hyperglycaemia and hyperuricaemia. In some patients fluid loss can be excessive; hypovolaemia and hypotension can occur, leading eventually to cardiovascular collapse.

Q11 Treatment of heart failure is directed towards improving symptoms. Harry should be counselled to give up smoking, as this has adverse effects on both the lung and cardiovascular system. To help his nausea and loss of appetite, he should be advised to eat small meals, low in salt and saturated fat but including fruit and vegetables. He should be advised to keep his body mass index (BMI) within the recommended limits (20–25), which may involve losing some weight.

Walking on level ground would be suitable exercise for patients like Harry; they are advised not to undertake strenuous exercise.

Harry's nights may be more comfortable if he uses additional pillows or elevates the head of his bed. He should be counselled to take his loop diuretic in the morning so that the diuresis occurs during the day and does not disturb his sleep at night.
As their condition deteriorates, most patients also require a vasodilator and some may also require a cardiotonic agent, such as digoxin, to improve cardiac output.

Key Points

- Tests to distinguish between angina and a myocardial infarction involve a full ECG and measurements of specific cardiac enzymes, such as creatine kinase MB and troponin I and T.
- Tissue fluid is formed in capillaries from the plasma, and its reabsorption involves a balance of hydrostatic and osmotic forces. When venous pressure rises, the balance is upset and excessive amounts of fluid remain in the tissues, causing oedema.
- In heart failure there is poor ventricular function and decreased stroke volume, which may lead to venous congestion and oedema in the abdominal organs and the lung.
- Thiazide diuretics are useful in early cardiac failure as they are relatively mild diuretics which can mobilize the oedema. Increasing the dosage of a thiazide above the recommended dose would elicit more side effects without markedly improving therapeutic effects. Changing to a more powerful agent from a different diuretic class is preferable if the patient's condition deteriorates.
- If the heart condition worsens, particularly when there is pulmonary oedema, a more powerful loop diuretic, such as furosemide or torasemide, is required to mobilize oedema fluid in the lung.

CASE STUDY 23 Grandpa's silence

Part 1

Q1 A stroke involves significant reduction in blood flow to a part of the brain. It can be caused either (i) by an embolus or by intravascular clotting, which blocks blood flow to an area (approximately 85% of strokes), or (ii) by haemorrhage from a ruptured blood vessel, which compresses the brain tissue (approximately 15% of strokes). Patients with extensive atherosclerosis are at risk of intravascular coagulation and blockage of cerebral blood flow, but a vessel can be blocked by a thrombus originating in another part of the circulation. This cause of stroke is common in elderly patients >60 years of age. Aneurysms which rupture suddenly are a more common cause of stroke in younger patients.

Q2 The brain forms about 2% of body weight but receives approximately 20% of the cardiac output and approximately 20% of the body's oxygen supply. Blood reaches the brain via two internal carotids and two vertebral arteries; the latter fuse inside the cranium to form the basilar artery. The carotid and basilar arteries are interconnected via the Circle of Willis, which forms a ring of blood vessels in the brain. This arrangement ensures that brain tissues can be supplied with blood from either the carotid or vertebral arteries and reduces the chances of an interrupted blood supply.
Flow to the brain tissue is precisely regulated by a process of autoregulation, according to local chemical conditions. Cerebral blood vessels dilate and so increase blood flow in response to decreased pH and arterial PO_2 and to increased arterial PCO_2, conditions associated with increased metabolic activity. The neurones are very sensitive to changes in cerebral blood flow; interruption of flow for a few seconds causes unconsciousness.

Q3

- The normal blood pH is 7.4 and the normal arterial PCO_2 is 40 mmHg (5.3 kPa)
- The acid-base disturbance is respiratory alkalosis
- Grandpa's high pH and low arterial PCO_2 is caused by hyperventilation (overbreathing).

This may be due to changes in the central respiratory centre which are due to the stroke itself or perhaps to mechanical ventilation given by paramedics when he stopped breathing at some point.

Q4 The kidneys are able to adjust blood pH by regulating reabsorption of bicarbonate and secreting or retaining H^+ according to the pH of the body.

When the blood is alkaline because of loss of CO_2, the kidney decreases both the reabsorption of bicarbonate and secretion of H^+. These two adjustments return blood pH towards normal (7.4).

Q5 Alkalosis can be caused by both metabolic and respiratory problems. Apart from hyperventilation, respiratory alkalosis can be produced by hypoxia, for example, when a person moves to high altitude with a reduced arterial PO_2, stimulation of respiration occurs via the peripheral chemoreceptors in the carotid and aortic bodies, which respond to the low arterial PO_2. Increased rate and depth of respiration causes an increased quantity of CO_2 to be lost from the body, and so pH rises.
Metabolic alkalosis can occur when there is excessive H^+ loss from the body, via loss of gastric contents in vomiting, or when a patient takes excessive quantities of antacid medication.

Q6 The respiratory system compensates for alkalosis by retaining CO_2. The central chemoreceptors in the medullary respiratory centres respond to the reduced H^+ by decreasing alveolar ventilation, which will increase blood arterial PCO_2 and return the pH to normal (7.4).

Part 2

Q7 Yes. The renal and respiratory compensations can normally rectify the changes in pH and blood gases, unless there are also problems within the lung or heart which limit normal gas exchange and cardiac output. Many stroke patients unfortunately also have concomitant conditions such as heart failure, atherosclerosis or diabetes, since strokes are more common in the elderly population.

Q8 Risk factors for stroke include: hypertension, hyperlipidaemia, hypercoagulability of blood, sluggish blood flow (e.g. following surgery or myocardial infarction) and atherosclerosis. All these predispose patients to stroke. Age could also be included here as the incidence of thromboembolism increases in those over 50 years of age.

Q9 No. Because some strokes are caused by cerebral haemorrhage rather than blockage of a cerebral vessel by a thrombus, haemorrhagic stroke must be ruled out by brain scans (e.g. a computed tomography, or CT, scan) before any agent acting on the coagulation mechanism is administered. Even in strokes known to be associated with thromboembolic events, the use of anticoagulants may increase the risk of converting the infarction of brain tissue into a haemorrhagic state. For this reason anticoagulants, if needed, are not normally administered until two weeks after the stroke has occurred.

Only one fibrinolytic agent has been recommended in the United Kingdom for treatment of acute stroke: alteplase, a plasminogen activator. This is used only under specialist supervision as there is a significant risk of intracranial haemorrhage.

Q10 Yes, aspirin is used in the secondary prevention of a further stroke. Aspirin inhibits the cyclooxygenase (COX) enzyme responsible for producing thromboxane in platelets. Thromboxane is involved in platelet aggregation, an early step in blood coagulation. A low daily dose of aspirin (75 mg) inhibits thromboxane production, preventing platelet aggregation and blood clotting in arteries.

Q11 The symptoms experienced by a patient when blood supply to part of the brain is disrupted depend on where the vascular blockage or bleeding occurs. The symptoms can include weakness or paralysis of muscle in one to four of the limbs, dizziness or fainting, numbness, visual disturbances or coma. Since the weakness in his limbs is on the right side of the body, Grandpa's brain damage is likely to be on the left side of his cerebral cortex. Nerve fibres descending from the motor cortex to innervate muscles in the limbs decussate (cross over) in the spinal cord or the medulla. If movement is affected on the right side of the body, cortical damage is likely to be in the left side of the brain.

In most people the area of cerebral cortex associated with speech is located in the left cortex, so both the speech difficulties and the right limb weakness are consistent with damage to the left side of Grandpa's cerebral cortex.

Q12 A stroke is normally treated conservatively, but it is important, during recovery, to reduce the chance of a further stroke. Low-dosage aspirin (75 mg daily) is recommended to reduce the risk of thrombus formation. If BP is high (hypertension), this is treated with an antihypertensive agent.

Patients are advised to stop smoking and should receive counselling help to achieve this. This is because nicotine replacement therapy, which has proved very helpful to many patients in giving up smoking, is not recommended for patients who have recently suffered a stroke.

Grandpa should also reduce his saturated fat consumption and substitute more healthy food for the double-cheese pizzas, burgers and fries! Reduction in saturated fat intake will improve his blood lipid profile somewhat, but Grandpa will probably also need a statin to reduce hypercholesterolaemia.

Many patients can recover at least part of their lost muscular activity following a stroke as the undamaged neurones can be induced to make new connections and take over some of the lost functions of the damaged area. Physiotherapy will be required to promote any improvement in muscle control and activity.

Key Points

- The brain uses a substantial proportion of body oxygen and there is a generous blood supply to the brain from the carotid and vertebral arteries. Interruption of brain blood flow for more than a very short time causes neuronal damage and ultimately cell death. Cerebral blood flow is normally controlled by autoregulation.
- Blockage of blood supply to part of the brain by a thrombus reduces its oxygen supply and damages neurones in the area, causing a stroke. Similar damage can be caused by a bleed into the nerve tissue. Risk factors for stroke include hypertension, hyperlipidaemia and clotting defects.
- It is important to ascertain whether the cause of a stroke is thromboembolic or haemorrhagic so that appropriate treatment can be initiated.
- If sites in the brain that control respiration are damaged, respiration and blood gas tensions will be disrupted. It is also possible that assisted ventilation is required by a stroke patient, and this can alter blood gas tensions temporarily. Renal and respiratory compensations rectify these changes during recovery.
- Aspirin is used in secondary prevention of stroke as it reduces platelet aggregation and the clotting tendency of blood.
- The outcome of stroke depends on the area of neurones involved and may include weakness or paralysis of muscles, numbness, visual or speech disturbances and coma.

CASE STUDY 24 The gardener who collapsed on his lawn

Part 1

Q1 An infarct is an area of dead tissue caused by interruption of blood flow. In a myocardial infarction (heart attack) caused by interruption of coronary blood flow, an area of cardiac muscle cells dies because of ischaemia. The patient may recover but the affected area of muscle may not survive: it may be without blood flow for a prolonged period or the coronary blood vessel may be permanently blocked. The affected cardiac muscle is eventually replaced by fibrous scar tissue.

Q2 A defibrillator administers a large direct current (DC) shock (starting at ~200 J) across the chest wall in order to depolarize the whole heart and stop the activity of the dysrhythmic areas, which are producing ectopic (abnormal) beats. It is hoped that this will allow the normal pacemaker, the SA node, to start again and generate the cardiac impulse in a normal rhythm: this is known as *sinus rhythm*.

Q3 Epinephrine is a non-selective adrenoceptor agonist, capable of stimulating both alpha- and beta-adrenoceptors (α- and β-adrenoceptors). When the heart is failing, epinephrine can stimulate contraction of cardiac muscle via β_1-receptors and so raise the cardiac output, without causing bronchoconstriction, as it simultaneously relaxes airways via its action on β_2-receptors. Epinephrine is used in emergency situations because, when given intravenously, it acts rapidly, in approximately 1 min. Sometimes the damaged heart develops an unsuitably slow beat, bradycardia, which needs to be increased to maintain an adequate circulation. The drug of choice for this action is atropine, which counteracts the effects of vagal slowing. Atropine is a cholinergic antagonist acting on cardiac muscarinic cholinoceptors.

Q4 Cardiac enzymes are released into the blood following heart muscle damage during a heart attack. Creatine kinase, particularly its MB isoenzyme, is one of the most specific of these enzymes, which reaches a peak 24 hours after infarction. It rises and then falls within the first 72 hours of the heart attack. Aspartate transaminase is also released, but levels of this enzyme can be raised in several other conditions, so it is less specific than creatine kinase MB. Troponin T is also specific for myocardial damage and is raised for approximately two weeks following infarction. Finding a high concentration of these enzymes in a patient's blood therefore supports the evidence obtained from the ECG and confirms that the patient has suffered a myocardial infarction.

Q5 The heart may be regarded as lying in the centre of an equilateral triangle: the apices of the triangle are marked by the two shoulders and the mid-line of the hips. The limbs act as linear conductors. Depolarization and repolarization of the cardiac cells causes small currents to flow through body fluids and tissues; this produces voltage changes at the body surface which can be measured from electrodes placed at specific points on the skin. The ECG amplifies and records these small voltage changes which occur at the body surface during each cardiac cycle.

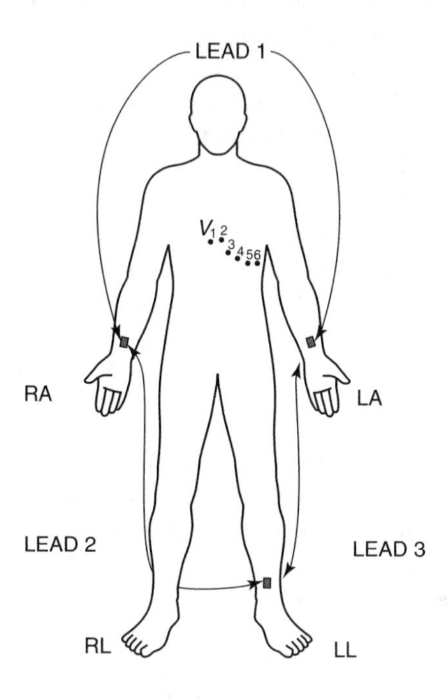

Six of the 12 leads of the ECG are placed directly on the chest wall. By international convention, these electrodes are placed in predetermined locations and record activity from sites directly over specific parts of the heart. The other six leads are associated with recordings from the limbs.

The conventional limb leads record potentials between two apices of the triangle. Lead 1 records potentials between the right arm and left arm, lead 2 records potentials between the right arm and left leg and lead 3 records potentials between the left arm and left leg. These are known as *bipolar leads*. In addition there are three unipolar leads attached to the limbs, which record potentials between the limb and a reference zero.

The normal ECG has a recognizable pattern which varies only a little from person to person. When electrical or rhythmic changes in the heart occur, specific changes can be seen in the ECG which can be used to aid diagnosis.

Q6 P QRS T

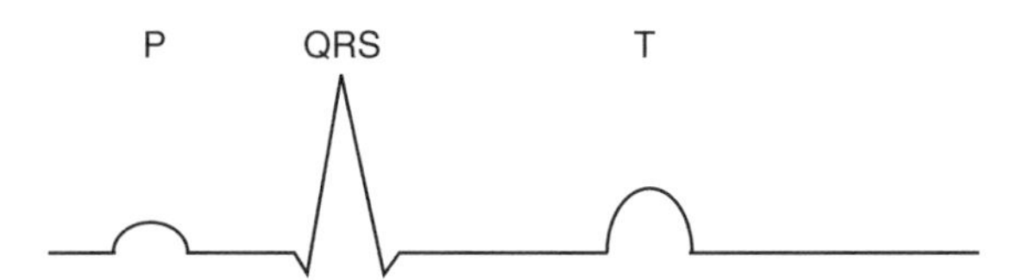

The first wave of the ECG is a small rounded P wave: this is followed by the spiky QRS complex and a larger rounded T wave.

Part 2

Q7 The purpose of using aspirin after a heart attack is to minimize the risk of blood clotting in the circulation. It is recommended (in the United Kingdom) for use in the long-term management of patients following myocardial infarction, unless there are contraindications to its usage in a particular patient.
Aspirin is a non-steroidal anti-inflammatory drug which inhibits the COX enzyme responsible for the production of a range of prostaglandins. The latter agents are involved in inflammation, control of body temperature, pain, platelet aggregation and many other body processes. Aspirin reduces fever and pain and stops platelets from aggregating (clumping), so preventing blood coagulation.

Q8 When a blood clot causes blockage to a coronary blood vessel, the cardiac muscle beyond the blockage starts to die. There is a short period following a heart attack when the muscle can be saved by dissolving the clot, so reinstating blood flow to the damaged region. Intravenous thrombolytic drugs, such as streptokinase, need to be given within 12 hours of the infarction, preferably in the first hour. They lyse clots by activating circulating plasminogen to plasmin. Plasmin degrades fibrin and so breaks up blood clots, allowing blood to flow through the coronary vessel once more. Streptokinase and tissue plasminogen activators have been shown to reduce mortality following acute myocardial infarction.

Q9 A patient who has recently received a head injury, or had surgery, or who has haemorrhaged recently may bleed excessively; in this situation, thrombolytics are contraindicated. These drugs are also unsuitable for patients who have previously experienced an adverse reaction to one of the drugs, such as an allergic reaction to streptokinase.

Q10 The word *angina* means pain; *angina pectoris* refers to pain in the chest.
The usual cause of pain arising from the heart is a deficiency of blood flow to cardiac muscle, which is usually caused by atheroma in coronary vessels, which restricts blood flow. Diminished blood flow to the heart muscle causes ischaemia and pain. A patient may have no problems at rest and show a

normal ECG trace when sitting or lying down, but he or she may complain that walking fast or climbing a hill, both of which require an increased delivery of oxygen to the cardiac muscle, brings on the angina.

Ischemic hearts show characteristic changes in the ECG, involving mainly the ST segment and T wave. These changes can be particularly prominent when the heart is stressed by exercise. Graded exercise, often using a treadmill, is used diagnostically to investigate possible myocardial ischaemia. Mild ischaemia is usually shown by a depression in the ST segment of the ECG.

Part 3

Q11 The failing heart cannot efficiently pump out the volume of blood which returns in the veins. Venous pressure therefore rises and tissues become oedematous. If the left heart fails, pulmonary venous pressure rises and oedema occurs in the lung.

When the patient stands up, there is a fluid shift towards the lower parts of the body and oedema is often first observed in the lower legs, particularly in the ankles. On lying down at night, there is a tendency for fluid to move from the lower body into the thorax, increasing pulmonary oedema. The decreased efficiency of gas exchange in the oedematous lung plus stimulation of sensory receptors in the lung causes dyspnoea. Characteristically, the patient wakes soon after retiring, feeling breathless or unable to breathe. Propping up the upper body with pillows, or raising the head of the bed, helps to keep any excess fluid in lower parts of the body, away from the lung. The patient is therefore more comfortable and is usually able to sleep.

Q12 Furosemide is a loop diuretic. The site of action of this drug in the nephron is the ascending limb of the loop of Henle. This tubule pumps sodium, potassium and chloride out of the filtered fluid into the medullary interstitial fluid, without accompanying water, producing a large osmotic gradient in the medulla. Since furosemide inhibits the activity of ion pumps in the ascending loop of Henle, the medullary osmotic gradient is diminished, less water can be reabsorbed in the collecting duct and a large amount of salt and water is excreted by the kidney. Loop diuretics are the most powerful of the diuretic drugs and can mobilize oedema fluid from the lung for excretion by the kidney. Reducing pulmonary oedema improves the gas exchange in the patient's lung and reduces dyspnoea.

Q13 The diuretic treatment should make Charlie more comfortable and improve his breathing. In addition he should be advised to reduce salt intake, try to keep his weight at a suitable level (BMI 20–25) and take gentle exercise, such as walking each day. He now needs someone else to do the heavy work, like digging, in the garden. Chronic heart failure is difficult to manage, as most patients' cardiac function gradually continues to decline. If systemic venous congestion has reduced his appetite, several small meals each day will be more suitable for him than one or two large ones.

Key Points

- Permanent interruption of blood flow to heart muscle results in cardiac cell death caused by ischaemia. Ischaemic damage causes release of cardiac enzymes into plasma, including creatine kinase MB and troponin I and T. The affected muscle is eventually replaced by scar tissue.
- Cardiac ischaemia may trigger abnormal electrical activity, causing fibrillation. Defibrillators deliver a large DC shock across the heart (cardioversion), to arrest abnormal activity and allow re-establishment of sinus rhythm.
- The 12-lead ECG consists of three bipolar and nine unipolar leads, six of which are placed directly on the chest wall. Damage to the heart muscle results in characteristic ECG changes.
- Aspirin at low doses is used in secondary prevention of heart attacks: aspirin reduces platelet aggregation, decreasing risk of thrombus formation and a further heart attack.
- In suitable patients a thrombolytic agent such as streptokinase may be given soon after a coronary occlusion to dissolve the thrombus and promote restoration of blood flow.
- Heart failure may follow a heart attack, as a result of cardiac muscle damage.
- Diuretics are often drugs of choice in mild heart failure.

CASE STUDY 25 Hannah's palpitations

Part 1

Q1 The mitral valve has two cusps and is located on the left side of the heart between the left (L) atrium and the L ventricle.

Q2 The AV valve on the right side of the heart is the tricuspid valve. It has three cusps and is located between the right (R) atrium and the R ventricle. In addition there are semilunar valves at the entrance to both the aorta and the pulmonary artery.

Q3 The red cell moves from the inferior vena cava to the right atrium, into the right ventricle and through the pulmonary artery into the arteries, capillaries and veins of the lung. From the lung it re-enters the heart via the pulmonary vein and travels through the L atrium and L ventricle before entering the aorta.

Q4 The cardiac impulse arises in the pacemaker tissue of the heart, the SA node. The nodal tissues of the heart–the SA and AV nodes in the right atrium – are spontaneously rhythmic. The impulse generated by the SA node spreads, rather like ripples on a pond, over the atria and reaches the AV node to excite it. From the AV node the impulse travels via the bundle of His along the Purkinje fibres, which are enlarged muscle cells with a high conduction velocity, to the ventricles. The cardiac impulse reaches the apex of the heart first and then spreads over the muscle of the two ventricles.

Q5 P QRS T

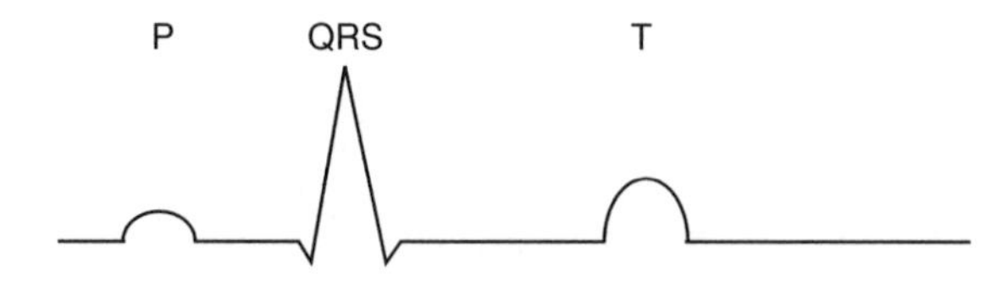

The first wave form of the standard ECG is the small rounded P wave associated with conduction of the cardiac impulse from the SA node over the atria; it represents atrial depolarization. This is followed by the spiky QRS complex, which represents a depolarization of the ventricles. The last major wave form is the rounded T wave, which represents ventricular repolarization: repolarization of the atria is hidden amongst the QRS complex.

Q6 The first heart sound, LUB, is associated with the closure of the mitral and tricuspid valves as the ventricles begin to contract. The second heart sound,

DUP, is due to closure of the semilunar valves in the outflow vessels from the ventricles, the aorta and pulmonary artery.

Q7 When valves in the heart do not close properly, there is backflow of blood from the ventricle into the atrium as the ventricle contracts. The backflow of blood causes turbulence and abnormal heart sounds, or murmurs.

Part 2

Q8 Depolarization of the atria normally gives rise to one P wave which precedes the QRS complex and a coordinated atrial contraction. In fibrillation the cardiac impulses arise abnormally and discharge at a very high rate ($>350\,\text{min}^{-1}$), producing a fast series of small, irregular waves before the QRS complex of the ECG. When this happens, the atria are unable to contract in a coordinated manner. Only occasional impulses can move through to the AV node to excite the ventricle, and ventricular rhythm becomes irregular. Patients become aware of the abnormal ventricular rhythm and usually describe the sensation as 'palpitations'.

Q9 Initial, compensatory responses to reduced cardiac output involve the sympathetic nervous system and release of adrenaline. The result is an increase in rate and force of cardiac contraction and arterial and venous constriction. In the short term this improves the circulation by increasing cardiac output. However, in the long term it is detrimental since vasoconstriction increases afterload, which, in turn, increases cardiac work and oxygen requirement, ultimately leading to cardiac failure.

Q10 Backflow of blood into the atrium from the L ventricle through the defective mitral valve increases the volume and pressure of blood in the L atrium, leading to atrial hypertrophy. Since some of the ventricular output returns to the atrium and does not enter the aorta, the ventricle needs to pump an increased volume of blood at each beat. This increases the work of the L ventricle, causing ventricular hypertrophy. The cardiac impulse may be conducted abnormally through the hypertrophied ventricle, leading to the development of ventricular dysrhythmias and possibly to cardiac failure.
In addition to ventricular changes, the increased volume and pressure of blood in the L atrium increases pressure in the pulmonary circulation, leading to pulmonary congestion. Pulmonary hypertension and oedema develop, predisposing patients to respiratory infections and dyspnoea during exercise. The patient suffers fatigue and weakness and sometimes produces blood-stained sputum (haemoptysis). These changes in the lung can account for Hannah's dyspnoea.

Q11 Digoxin is a cardiotonic agent. It acts by increasing the concentration of calcium in cardiac muscle, so increasing the force of contraction.

The increase in calcium is indirect. Digoxin blocks the active sodium-potassium pump in cardiac cell membranes, by competing for the site at which K^+ acts. The increased sodium content of the cell, which occurs as a result of blocking the pump, decreases the concentration gradient for sodium to enter the cell. Calcium ions usually exit the cell via a passive exchange process, as the sodium ions enter. So any decrease in sodium entry reduces calcium exit, and the concentration of calcium in the muscle cell increases.

Digoxin has other, indirect, effects on the heart. Acting via the vagal centres in the brain, digoxin causes cardiac slowing (bradycardia) and slower impulse conduction. The net effect is a decreased heart rate, allowing more time for ventricular filling and a slower rate of conduction through the AV node. This outcome is particularly useful in atrial fibrillation, which is Hannah's problem.

Q12 Digoxin has a low or narrow therapeutic index, which means that the therapeutic concentration of the drug is close to a concentration that elicits toxicity. A particular problem with digoxin is the production of abnormal cardiac rhythms.

The sodium pump normally creates a small potential across cardiac cell membranes; when digoxin blocks this pump, there is some depolarization of the cell. The heart then becomes more excitable and abnormal rhythms or ectopic beats may occur. Some patients also experience gastrointestinal disturbances, such as anorexia, nausea or vomiting. When blood concentration of digoxin is high, there may also be CNS effects, which can include confusion and visual disturbances.

Q13 Arterial emboli, which can block blood vessels and cause ischaemia or infarction in the tissues they affect, tend to originate in the left heart and are associated with valvular disease and dysrhythmias. Mitral stenosis is associated with abnormal atrial rhythm, particularly atrial fibrillation. Fibrillation and other rhythm abnormalities in the atria favour blood coagulation, resulting in production of thromboemboli which can move to distant parts of the circulation, such as the cerebral circulation. Thrombi could also form on surfaces of valves distorted by calcification and other abnormalities. In view of the risks of thromboembolism, it is usual to provide anticoagulant therapy to patients with mitral valve problems and atrial fibrillation.

Q14 Warfarin is an orally active anticoagulant used in the treatment of valvular disease and atrial fibrillation. It is structurally similar to vitamin K, a compound which is required for the synthesis of prothrombin and several other clotting factors in the liver. Warfarin interferes with the actions of vitamin K and so reduces the risk of blood clotting. When taken by mouth, its effect is not immediate and it takes several days to achieve the maximal clinical effect.

Q15 Patients' responses to warfarin therapy vary; the extent of the anticoagulant effect is impossible to predict accurately. If the dose is too high for that particular individual, excessive bleeding may occur. Because the dosage is so critical, it is usual to monitor a patient's prothrombin time daily, or on alternate days, when therapy is started; this is a standard test of the time required for a key stage in blood coagulation to occur.

Patients prescribed warfarin are usually given advice and an anticoagulant treatment booklet with further information on their therapy, when the drug is first dispensed.

Key Points

- There are two sets of heart valves: the AV valves between atria and ventricles and the semilunar valves at the entrance to the pulmonary artery and aorta. Closure of these valves produces the first and second heart sound. Valves may become distorted and incompetent following infections such as rheumatic fever, leading to backflow of blood and abnormal heart sounds, or murmurs.
- The cardiac impulse starts in the SA node in the right atrium, spreads over the atria to excite the AV node and down the bundle of His to excite the ventricles.
- Atrial depolarization results in the P wave of the ECG, the QRS complex denotes ventricular depolarization and the T wave represents ventricular repolarization.
- Atrial fibrillation may promote blood coagulation; the clots produced can move to obstruct arteries at distant sites such as the coronary or cerebral circulation.
- Digoxin is a cardiotonic agent with a narrow therapeutic index which is used to treat atrial fibrillation since it not only increases cardiac contractility but also acts on vagal centres in the brain and beneficially slows the heart. A slow heart facilitates coronary perfusion and ventricular filling, which improves cardiac output.
- Warfarin is an oral anticoagulant which interferes with the actions of vitamin K to reduce production of blood-clotting factors. Dosage of warfarin is critical and patients' clotting times are monitored to allow adjustment of dosage to their particular requirements.

5

Respiratory disorders

CASE STUDY 26 Moving to England

Q1 An ordinary cold is not likely to account for Mrs Smythe's symptoms. Colds are very unlikely to persist for several months and the eyes are not usually much affected. Colds are normally self-limiting and last for approximately five to seven days. An alternative diagnosis, which accounts for Mrs Smythe's symptoms and their duration, is hay fever.

Clinical Physiology and Pharmacology Farideh Javid and Janice McCurrie

Q2 *Hay fever* (a common term for seasonal allergic rhinitis) is an allergic reaction induced by an immunoglobulin-mediated inflammatory response of the nasal mucosa to allergens, particularly pollen. There is inflammation of the upper respiratory tract, eyes and often the paranasal sinuses and throat. The major symptoms are sneezing, itchiness and increased secretion from the nose (rhinorrhoea, or runny nose) together with itchy, red, watery eyes. Other symptoms can include headache and changes in the patient's ability to smell. The symptoms can be very troublesome, interrupting daily activities and disrupting leisure and sporting pastimes.

Q3 Fexofenadine is an antihistamine. This agent is also of use in urticaria.

Q4 Antihistamines are effective in managing many of the troublesome symptoms of allergic rhinitis. Histamine is a neurotransmitter and a mediator of type 1 hypersensitivity reactions, such as urticaria and hay fever. There are several types of histamine receptors and these allergic conditions can be treated with H_1 receptor antagonists, such as promethazine, chlorphenamine and fexofenadine. First-generation antihistamines, such as promethazine, cause sedation and possess side effects associated with actions on muscarinic receptors. Fexofenadine is a newer drug with a longer duration of action, which does not sedate the patient.

Q5 Histamine is released in:

(1) inflammation

(2) allergic reactions

(3) tissue damage, for example in response to venoms (bee stings).

Q6

- H_1 receptors–located in the gastrointestinal (GI) tract, mediate GI contraction.
- H_2 receptors–located in the GI tract and cardiovascular system, mediate gastric secretion and cardiac stimulation
- H_3 receptors–located in the central nervous system (CNS) (pre-terminal and autoreceptors) may be involved in movement control.

Q7 Fexofenadine is a metabolite of another antihistamine, terfenadine, but has little or no cardiac toxicity. The development of sedation and antimuscarinic effects are limited since fexofenadine cannot easily cross the blood–brain barrier (only a very small amount can cross this barrier). The recommended adult dosage is 120 mg once daily. It is also recommended for children above 12 years of age.

Q8 Examples include: famotidine, ranitidine, nizatidine and cimetidine. They prevent food, histamine and acetylcholine-induced gastric-acid secretion. They are used to heal gastric and duodenal ulcers and in gastro-oesophageal reflux disease.

Key Points

- Hay fever, or allergic rhinitis, is an allergic reaction induced by an immunoglobulin-mediated inflammatory response of the nasal mucosa to allergens, particularly pollen.
- This condition causes an inflammation of the upper respiratory tract, eyes and often the paranasal sinuses and throat.
- Major symptoms are sneezing, itchiness and increased secretion from the nose (rhinorrhoea, or runny nose) together with itchy, red, watery eyes. Other symptoms can include headache and changes in the patient's ability to smell.
- H_1 receptor antagonists, including promethazine, chlorphenamine and fexofenadine, are effective in managing many of the troublesome symptoms.

CASE STUDY 27 The sneezing boy

Q1 Allergic rhinitis.

Q2 Perennial and seasonal allergic rhinitis affects many individuals and can cause serious complications, such as otitis media and chronic sinusitis. The symptoms of allergic rhinitis can be caused by house dust mites, pollens, moulds and other allergens.

Q3 A type 1 hypersensitivity reaction is responsible for the development of the allergy. The symptoms are due to the effects of mast cell degranulation with the release of histamine. Mast cells are located in the nasal passages and the nasal mucosa is sensitive to the effects of histamine released from these cells, leading to inflammation of the mucous membranes of the nose. The inflammation is associated with oedema and swelling, vasodilation and an increase in the secretion of mucus. The mucous membrane of other sections of the respiratory tract (accessory sinuses, nasopharynx, and upper and lower respiratory tract) will also be affected by the allergic reaction.

Q4 Perennial allergic rhinitis can be treated with antihistamines and corticosteroids.

Q5 Azelastine hydrochloride is an antihistamine, an H_1 receptor antagonist which is available as a nasal spray.

Q6 Antihistamines should be used with caution in patients with asthma. This is due to a reduction in expectoration following the drying effect of the drugs, which may thicken the bronchial and bronchiolar secretions.

Q7 An alternative medication could be the use of topical nasal corticosteroids, such as beclometasone or budesonide, administered as a nasal spray: cromoglicate may also be used. The mechanism of cromoglicate is poorly understood; it may stabilize the mast cells to reduce degranulation and histamine release. It is useful in the prophylaxis of both asthma and allergic rhinitis. The topical antihistamines are less effective than topical corticosteroids, but more effective than cromoglicate. Cromoglicate, however, is the first choice in children <12 years of age.

Key Points

- Perennial and seasonal allergic rhinitis are type 1 hypersensitivity reactions to an allergen.
- The symptoms are due to the effects of mast cell degranulation. The effects can cause serious complications, such as otitis media and chronic sinusitis.
- Allergens which cause these symptoms include house dust mites, pollens and moulds.
- Treatment of allergic rhinitis includes antihistamines, H_1 receptor antagonists, such as axelastine, and corticosteroids, such as beclometasone or budesonide. However, cromoglicate is the first choice for children.

CASE STUDY 28 Mandy's sleepover

Q1 The most commonly used 'reliever' in asthma therapy is a short-acting bronchodilator, such as the beta-2-agonists (β_2-agonists) salbutamol or terbutaline. These are safe and effective agents for mild to moderate symptoms and are taken directly into the respiratory tract via an inhaler device.

If patients need to use the reliever more than three times a week, they are usually also prescribed a 'preventer' inhaler containing a corticosteroid, such as beclometasone diproprionate, budesonide or fluticasone proprionate. Corticosteroids decrease airway inflammation, reducing airway oedema and mucus production. When used regularly they are prophylactic and reduce the frequency of asthma.

Q2 Asthma involves reversible narrowing of small airways in the lung. In acute asthma the smooth muscle surrounding the bronchi and bronchioles contracts, narrowing the lumen. Concurrently, airway mucous membranes become inflamed and oedematous and mucus secretion is increased; these changes cause further narrowing and obstruct airflow. Expiration is more severely affected than inspiration since expiration is passive and involves the recoil of lung structures stretched by the active inspiratory process. Patients have difficulty in moving air through their airways, which causes breathlessness, or dyspnoea. Wheezing is caused by turbulent and restricted airflow through the airways, and coughing is triggered by irritation of lung sensory receptors.

Q3 Risk factors for asthma include a genetic susceptibility and infection, for example a viral respiratory illness. There is evidence for a strong genetic component involving a number of genes rather than a single abnormality, or an 'asthma gene'; in this case Mandy's father and brother are both asthmatic. Some asthma attacks can be triggered by exercise. There may be many environmental triggers in Jane's house, including allergens from the pets' hair and skin cells and their urinary proteins, as well as cigarette smoke and house dust mites. Inhalation of such allergens in susceptible individuals leads to degranulation of pulmonary mast cells with release of mediators which cause mucosal inflammation, oedema and bronchospasm. Airway resistance is increased and wheezing, dyspnoea and coughing occur.

Part 2

Q4 Asthma affects expiration more than inspiration, and so tests of expiration are useful in determining the severity of the condition and the response to therapy.

Forced vital capacity (FVC) measures the maximum volume of air expelled from the lung in a single forced expiration: there is no time limit. Forced expiratory volume in one second (FEV_1) measures the volume of air which can be expelled from the lung in one second. In a normal individual 80% of the vital capacity can be expired in one second, but patients with obstructive disease have difficulty in emptying the lung and this value is significantly reduced.

The FEV_1/FVC ratio is a useful single measure of expiratory function. In a normal individual this ratio is likely to be 0.8 or more.

A ratio of <0.7 indicates some obstruction to expiratory airflow.

In Mandy's case: FEV_1/FVC ratio $= 950/2300 = 0.41$.

This ratio indicates obstructive disease.

Q5 Measurement of peak expiratory flow: this is a simple measure of expiratory function. The peak flow meter measures the velocity of expired airflow and is suitable for both adults and children. The patient breathes out a short blast of air, as fast as possible, into the device. Normal individuals can achieve airflow velocity of 450–650 l min^{-1}. The peak flow meter is a cheap device which is used by patients at home to monitor their asthma. If a patient's peak flow diminishes below a certain level which has been set by their nurse practitioner or family doctor, they can adjust their own treatment, within specified limits, and control their condition better.

Q6 Salbutamol is a selective beta-2-adrenoceptor (β_2-adrenoceptor) agonist which is effective in relieving mild to moderate bronchoconstriction.

Inhalation of salbutamol induces bronchodilation by acting on β_2-receptors on bronchial smooth muscle; this lasts for approximately three to five hours. It also inhibits mediator release and improves the clearance of mucus from the lung. Stimulation of the β_2-receptor increases the cellular concentration of cyclic adenosine monophosphate cAMP and activates a protein kinase. This kinase in turn inactivates myosin-light-chain kinase, an enzyme necessary for contraction in smooth muscle, and so relaxes bronchial smooth muscle.

Q7 Nebulizers convert a solution or suspension of drug into an aerosol which is administered by inhalation. The aerosol is able to carry a higher concentration of drug deep into the lungs than the dry-powder type of inhaler used normally by asthmatic patients. Nebulizers are useful when a patient has a more severe episode of asthma than usual, which is not relieved by their normal inhaler.

Good coordination is required in the use of metered dry-powder inhalers; using a nebulizer has the advantage that no coordination in drug delivery is needed by the patient. This is important if the asthmatic condition is severe and the patient is very young, or very anxious or confused.

It would be expected to fully reverse Mandy's bronchoconstriction.

Q8 Because Mandy's airways were constricted and obstructed, she was not able to empty her lung effectively during expiration and CO_2 was retained. Increased arterial CO_2 decreases arterial pH.

Q9 Yes. Her ability to tell staff about her usual medication shows that, although her asthma was moderately severe, it was not life-threatening. In very severe asthma patients cannot complete a sentence in one breath or may be too breathless to talk at all.

Q10 Other bronchodilator agents include nebulized ipratropium. Ipratropium is a muscarinic receptor antagonist that helps to relax bronchial smooth muscle which has contracted via parasympathetic stimulation. The xanthines theophylline and aminophylline (theophylline ethylenediamine) are alternative bronchodilator agents. These agents may act as phosphodiesterase inhibitors and, although they have been used as bronchodilators for many years, adverse CNS, GI and cardiovascular effects may limit their usefulness.

Q11 When dry-powder metered-dose inhalers are used, there is some deposition of the drug dose in the mouth and pharynx. These inhaler devices need good coordination between activation of the device and the inhalation of the drug: very old, young or anxious/confused patients may not be able to coordinate well. Spacer devices both eliminate the requirement for good coordination and reduce the deposition of drug in the oropharynx. More of the drug is able to enter the lung and so the therapeutic effect of the agent is optimized. Spacers are particularly useful for very young children with asthma.

Q12 Children who need more than occasional relief of bronchoconstriction are usually prescribed a standard corticosteroid inhaler as prophylaxis.

There is some evidence that children under five years of age obtain benefit from use of nedocromil sodium or sodium cromoglicate. These agents are used only in prophylaxis: cromoglicate is not a bronchodilator and cannot be used to treat acute episodes of asthma. Its action is not well understood but the prophylactic effect appears to be partly due to stabilization of mast cells, which reduces release of histamine and other mediators so that hyperactive bronchial muscle is less responsive to environmental triggers.

Other, recent additions to prophylaxis in asthma therapy include the leukotriene receptor antagonist montelukast. This drug is taken as a tablet and blocks the actions of cysteinyl leukotrienes in the airways. The latter are products of the lipoxygenase pathway which cause bronchoconstriction and inflammation. It is no more effective than standard corticosteroids in the prophylaxis of asthma, but there is some evidence that when given together with a steroid there may be a beneficial additive effect.

Key Points

- Asthma involves reversible bronchoconstriction, which particularly affects expiration.
- Patients with asthma are usually treated with a 'reliever', usually a short-acting β_2-agonist, and a 'preventer' inhaler containing a corticosteroid.
- Children may benefit from asthma prophylaxis using sodium cromoglicate or nedocromil sodium.
- Risk factors for asthma include genetic susceptibility, infection and exposure to triggers such as cold air, animal products and house dust mites.
- Respiratory function tests of particular use in asthma include peak expiratory flow and FEV_1.
- Nebulizers are useful in treating severe asthma as they administer bronchodilator drugs as an aerosol and, unlike dry-powder inhalers, require no coordination by the patient in their use.
- Spacer devices are useful to deliver drugs into the respiratory tract of young children with asthma. They reduce the deposition of bronchodilator drugs in the pharynx and require little coordination by the patient to deliver the required dose.

CASE STUDY 29 Bob and Bill's breathing problems

Part 1 Bob's problems

Q1 *Dyspnoea* is the subjective sensation of discomfort in breathing. It does not describe a pain, but patients may refer to it as 'shortness of breath' or 'breathlessness'. Many different lung conditions can give rise to this sensation. *Tachypnoea* is a term meaning rapid breathing. The normal breathing rate for an adult is between 12 and 16 breaths per minute and the tidal volume is normally 400–500 ml. In some pulmonary diseases patients show tachypnoea, a more rapid, but shallow, breathing.

Q2 There is a wide range of respiratory function tests available; many can be performed using spirometry or simple equipment such as the peak flow meter. The tests are used to aid the diagnosis of the respiratory disorder present, to follow the course of the disease, which may be recurrent or progressive, and to monitor the effects of therapy. In addition, there are specific occupational lung diseases, for which patients who have a respiratory disability, because of adverse conditions at the workplace, may claim some financial compensation. Their compensation depends on the extent of their respiratory disability.

Q3 Bob's FEV_1/FVC ratio $= 2500/2700 = 0.93$.
Both of the parameters are reduced compared to normal values, but since both are diminished similarly the ratio is normal.

Q4 The diminished vital capacity and forced expiration in one second, together with a normal FEV_1/FVC ratio, is characteristic of restrictive pulmonary disease.
It is unlikely that a bronchodilator will be useful to Bob as his problem involves a change in the substance of the lung tissue, which restricts inflation and deflation, but there is usually little bronchoconstriction.

Q5 Restrictive lung disease reduces lung capacity and results in rapid, shallow breathing. This type of respiration tends to wash CO_2 out of the lung and may result in an increase in blood pH. Gas exchange in alveoli is reduced or inadequate because of the poor expansion of lung tissue, so the arterial PO_2 of arterial blood may also be rather low.

Q6 Yes. Bob has been exposed to dusts from the sand-blasting of buildings, and possibly to asbestos in insulation materials. Either of these could lead to an occupational lung disease. Basically, the inhaled dusts irritate the lung and set up chronic inflammatory changes. In response, collagen is deposited, the lung tissues become more fibrous, elasticity is lost and a restrictive lung condition develops.

Part 2 Bill's problems

Q7 Bill's FEV_1/FVC ratio $= 1000/2700 = 0.37$. This is much lower than the normal ration of 0.8 and shows that he has a chronic obstructive pulmonary disease (COPD).

Q8 The two major types of COPD are chronic bronchitis and emphysema. It is not possible to determine which of these two conditions is responsible for the problems of this patient from the information given, but they often coexist. Chronic bronchitis is characterized by recurrent chest infections with a productive cough and sputum production for at least three months in two or more consecutive years. In chronic bronchitis there is hypertrophy of the mucous glands in the airways and production of a thick, tenacious mucus that is difficult to remove from the lung and which easily becomes infected. The incidence of bronchitis is increased in smokers.

Emphysema is defined as a condition in which patients have permanent enlargement of airspaces distal to the terminal bronchioles with destruction of their walls. Loss of alveoli and bronchioles in emphysema is permanent and irreversible. In most patients emphysema is initiated by inhalation of inflammatory oxidants, such as those in cigarette smoke; smoke also reduces the activity of cilia so that mucus clearance is reduced. Inflammatory mediators and enzymes, such as elastases, are released in the lung and begin the destruction of lung tissues. This is normally opposed by α_1-antitrypsin, which inhibits the activity of elastases. When there is a deficiency of α_1-antitrypsin, patients are predisposed to emphysema, even if they are non-smokers. Eventually, large airspaces are created in the lung, the surface area for gas exchange is greatly diminished and abnormality in arterial gas composition and pH results.

Q9 The arterial blood gases are abnormal because of reduced surface area for diffusion, which leads to poor gas exchange. The arterial PCO_2 is higher than normal because of a retention of CO_2, and arterial PO_2 is lower than normal. A high arterial PCO_2 results in acidosis (pH 7.3, instead of the normal 7.4).

Q10 Possibly. Allergens from the birds' feathers and excreta are inhaled and can cause inflammatory changes in the human lung, leading to bronchoconstriction. It is also possible that the tatty parrot could be a source of the viral disease psittacosis, which normally affects birds but can infect the human lung when people come into close contact with infected birds. The symptoms include fever, shortness of breath and cough. This disease is common in imported parakeets.

Q11 Smoking is the most important risk factor in the development of COPD and is considered the single greatest cause of preventable illness. Stopping smoking both reduces the risk of heart disease and decreases mortality from COPD.

Although lung elasticity naturally decreases with age, smoking increases the rate of decline and giving up smoking slows this deterioration.
Motivation to quit smoking is the most important factor in smoking cessation. Bill will require considerable help and support to maintain his motivation. There are several types of nicotine-replacement therapy which will help: nicotine patches, gum, nasal sprays and inhalers are available. If these are not suitable, there are non-nicotine methods of cessation including acupuncture, hypnosis and alternative drugs such as amfebutamone (bupropium), which was originally developed as an antidepressant. Ideally, Bill should be given the information on the various methods of smoking cessation and be supported to make his own choice of the method he prefers.

Q12 It is recommended that a trial of a short-acting beta-2-agonist (β_2-agonist) inhaler be made for a few weeks as some COPD patients do benefit from bronchodilation. Although his doctor has prescribed a bronchodilator previously, it may be useful for Bill to try this again. There should also be a trial of a corticosteroid inhaler, as this diminishes the inflammatory component of COPD. If there is no appreciable benefit after four weeks, the steroid should be discontinued.

Key Points

- Lung volumes are changed differently by restrictive and obstructive disease. In restrictive disease most volumes and capacities are decreased to the same extent and the ratio of FEV_1/FVC is within the normal range (> 0.8). In obstructive disease FEV_1 is greatly reduced and the FEV_1/FVC ratio is decreased (< 0.7).
- Obstructive lung disease is commonly associated with smoking or prolonged exposure to industrial smokes and fumes. The destruction of lung tissue in emphysema is permanent and irreversible and development of the condition is linked to deficiency of alpha-1-antitrypsin (α_1-antitrypsin).
- There may be an inflammatory component in obstructive lung disease, and a trial of corticosteroids and bronchodilators is recommended.
- Patients with obstructive lung disease who smoke should be helped to quit smoking in order to reduce their risk of heart disease and to decrease mortality.

CASE STUDY 30 A punctured chest

Part 1

Q1 *Dyspnoea* is defined as: shortness of breath and difficulty in breathing, usually caused by lung or heart disease.
Atelectasis is defined as: a condition in which part of a lung or a whole lung collapses and alveoli completely deflate.
Pneumothorax is defined as: the presence of air or gas in the pleural space caused by rupture of the pleura.
Cyanosis is defined as: a condition in which the skin and mucous membranes appear blue because of the lack of oxygenated haemoglobin. Desaturated (reduced) haemoglobin has a bluish colour, which is seen most easily in the lips; it is observed when arterial PO_2 is reduced.

Q2 Air can enter the lung via the nose or mouth, passing through the glottis and entering the trachea. From the trachea, air moves into two bronchi and flows

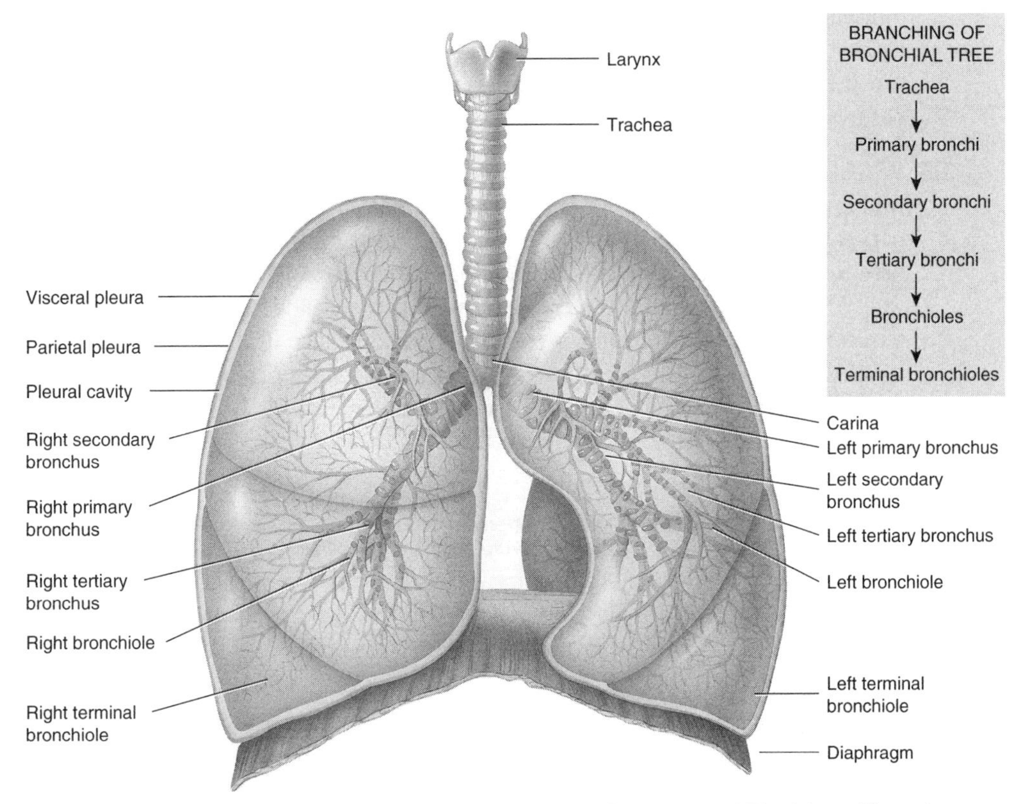

Anterior view. From Tortora and Derrickson *Principles of Anatomy and Physiology, Eleventh Edition 2006.* Reproduced with permission of John Wiley & Sons, Inc.

through their many branches to reach the alveoli. In inspiration the volume of the thorax is increased by contraction of the diaphragm and external intercostal muscles; this lowers intrathoracic pressure and draws air into the thorax. The outer surface of the lung and the inner surface of the thoracic wall are in contact with the pleural membranes. As the chest wall expands the pleura are forced to follow as they cannot normally be pulled apart; the lung also follows and expands as it cannot normally be separated from the pleura. This arrangement allows the lung to inflate when the thoracic volume increases.

Q3 Expiration in quiet breathing is passive. When inspiration ceases and the intercostal muscles and diaphragm relax, the volume of the thorax diminishes and the elastic tissues of the lung recoil. This recoil is sufficient to move the normal expiratory volume of air out of the lung.

Q4 The pleura are serous membranes: one layer (the visceral pleura) firmly adheres to the surface of the lung and the other (the parietal pleura) adheres to the inner surface of the thoracic wall and diaphragm. The two pleural membranes lie very close together, separated only by a thin film of fluid. This lubricates the pleural surface, allowing the two layers to smoothly slide over each other as the thoracic wall moves.
The lung tissue is arranged in lobes: the right side has three lobes and the left has two lobes. Each lung is surrounded by separate pleural membranes.

Q5 When air enters the pleural cavity, either from the outside when the chest wall is punctured or from the lung itself if alveoli rupture, the visceral and parietal pleura become separated. The consequence of the introduction of air between the pleura is that the lung does not adhere to the pleura and thoracic wall when thoracic volume increases. Instead, the elastic fibres of the lung tissue and the surface tension of the air–water interface in the alveoli cause lung tissue to recoil and eventually collapse. To reinflate a collapsed lung the hole in the chest wall must be closed, a small tube placed in the pleural cavity and suction applied to remove the air from the cavity.

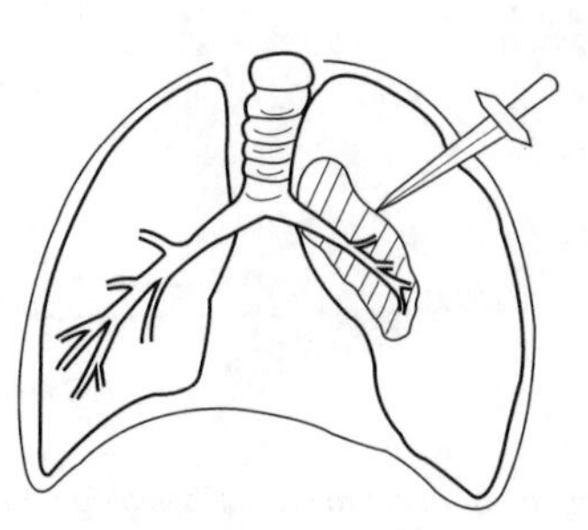

Pneumothorax

Q6 Each lung is enclosed in a pleural membrane; so the left pleural compartment is completely separate from the right compartment, which was penetrated by a nail. The pleura on the left side have not been disrupted, and so the movement of the left lung in inspiration and expiration was not affected by an injury on the right side.

Q7 Brad's arterial PO_2 is lower than normal and the arterial PCO_2 is higher than normal. The arterial PO_2 is reduced mainly because the right lung has collapsed: ventilation has been reduced and the surface area for gas exchange and transfer of oxygen into the blood is greatly diminished. arterial PCO_2 is raised because of reduced ventilation and decreased elimination of CO_2. Because CO_2 is being retained, Brad is experiencing respiratory acidosis and his pH is low. Since Brad is a heavy smoker (>40 cigarettes each day), it is possible that he already has some lung dysfunction. This adds to the blood gas abnormality caused by the pneumothorax.

Part 2

Q8 Brad's FEV_1 and peak expiratory flow rate are a little lower than expected for a person of his age and height. The FVC is just a little higher than expected. The low peak flow and FEV_1 suggests that he may now be experiencing some obstruction to expiration.

Q9 The ratio FEV_1/FVC is helpful in distinguishing between restrictive and obstructive lung conditions. In restrictive diseases most of the pulmonary function tests show lower values than normal, but the FEV_1/FVC ratio is normal (>0.7). In obstructive disease there is a problem with airflow in expiration, air is retained in the lung and total lung capacity gradually increases. In an obstructive type of disease only the tests of expiratory function reveal low values; it is considered to be indicative of an obstructive condition when the ratio of FEV_1/FVC is <0.7. Brad's ratio: $FEV_1/FVC = 3.4\,l/4.8\,l = 0.68$. Since the ratio is <0.7, there is evidence that Brad is developing an obstructive condition. This is not yet marked, and stopping smoking is likely to slow the deterioration of his lung function significantly.

Q10 It is possible that Brad's low arterial PO_2 is related to his smoking habit. The particulates and chemical constituents in cigarette smoke irritate lung tissues, causing inflammation and increased mucus production. The inflammatory change and accumulating mucus can cause obstruction to expiratory airflow and add to the mismatch of ventilation and perfusion, which is a characteristic of obstructive lung disease. So oxygenation of blood decreases and arterial PO_2 is lower than normal.

Q11 Carbon monoxide (CO) combines with haemoglobin at the same place as oxygen, producing carboxyhaemoglobin. Since its affinity for haemoglobin is approximately 250 times greater than oxygen, the combination of CO with haemoglobin is difficult to reverse. Burning any organic substance in air can produce CO, burning the tobacco in a cigarette is no exception. Smokers inhale CO with the cigarette smoke and CO combines cumulatively and irreversibly with their haemoglobin, gradually decreasing the ability of their blood to carry oxygen. The blood of smokers may contain 10% or more of CO, and this both reduces the oxygen-carrying capacity of their blood and impairs cognitive function.

Q12 There are hundreds of chemicals in cigarette smoke which can potentially irritate lung tissue, causing inflammatory change and increased mucus production. The smoke decreases the activity of cilia which normally beat upwards, towards the pharynx, to clear secretions and debris from the lung. Accumulated mucus irritates the sensory receptors in the lung, stimulating the cough reflex and production of sputum. When the inflammatory changes persist, the elastic tissue in the lung may be gradually replaced by fibrous tissue, decreasing the recoil which normally facilitates expiration and contributing to the development of obstructive disease.

Approximately 60 of the components of cigarette smoke are known to be carcinogens, and their accumulation in the lung is an important risk factor in the development of lung cancer.

Key Points

- Inspiration is an active process involving the diaphragm and external intercostal muscles. Expiration is normally passive, because of relaxation of these muscles and recoil of lung tissue.
- Pleural membranes cover the lung and must be intact to allow the lung to inflate. If the chest wall is injured and the pleura punctured, air enters the pleural cavity (pneumothorax) and the lung collapses.
- Lung collapse (atelectasis) reduces FEV_1 and is likely to affect blood gas tensions, reducing arterial PO_2.
- CO has a very high affinity for haemoglobin. Smoking cigarettes results in repeated inhalation of CO and causes carboxyhaemoglobin to accumulate in the patient's blood, reducing its oxygen-carrying capacity.

CASE STUDY 31 Carmen's repeated respiratory infections

Part 1

Q1 Sweat glands are innervated by the sympathetic division of the autonomic nervous system. The postganglionic transmitter is, however, acetylcholine.

Q2 In the sweat test, sweating is induced by passing a weak electric current across an area of skin treated with the secretory stimulant pilocarpine. The current enhances the ability of pilocarpine to penetrate skin and so local secretion of sweat is induced. When this test was performed on Carmen, the diagnosis was confirmed: her sweat was found to contain 100 mmol l^{-1} chloride, normally this is expected to be <60 mmol l^{-1}.

Q3 Pilocarpine is a partial agonist at muscarinic cholinoceptors; when applied to an area of skin, it stimulates muscarinic receptors on the sweat glands to cause local sweating. Pilocarpine shows some selective activity in that it stimulates secretion from exocrine glands such as sweat, salivary and bronchial glands more strongly than smooth muscle of the gastrointestinal and urinary tract. It is one of two muscarinic agonists in clinical use. Apart from its use in diagnosis of cystic fibrosis, by stimulating sweat secretion, pilocarpine is used in the management of glaucoma to contract the ciliary and iris muscles in the eye, so reducing intraocular pressure.

Q4 The cough reflex is protective as it removes secretions and debris from the airways. Coughing may be loose and produce sputum or be dry and irritating. It is usually initiated by irritation of sensory receptors in the tracheobronchial tree; the stimulus may be mechanical, chemical or inflammatory.
Cough begins with a deep inspiration, followed by a forced expiration against a closed glottis. The pressure in the chest rapidly rises and when the glottis opens suddenly an explosive outflow of air is produced.
The cough reflex helps to expel irritants, mucus and infected material from the lung and keeps the airways clear. It is the most common symptom of respiratory disease and smokers often have a chronic cough because of irritation from inhaled cigarette smoke. Although primarily protective, cough can be exhausting for patients with severe lung disease.

Q5 Since a major problem in cystic fibrosis is production of very viscous mucus which is frequently infected, the cough reflex should not be completely suppressed. Failure to cough and at least partially clear the airways would cause secretions to be retained, forming obstruction and a focus for further infection.

Q6 Abnormal salt and water transport in airway epithelia results in production of a thick, sticky mucus which is difficult to move out of the lung by ciliary action. The patient may not be able to mobilize this mucus by coughing and the clearance of mucus often requires some mechanical help. In physical therapy, the patient's chest is vibrated or percussed vigorously with cupped hands; this vibration of the chest helps to dislodge the tenacious mucus and move it towards the pharynx to be expelled.

Q7 Three classes of drug have acute bronchodilator activity:
Beta-2-agonists β_2-agonists) such as salbutamol (short-acting) or salmeterol (longer-acting) are effective bronchial muscle relaxants administered by inhalation.
Muscarinic antagonists, such as ipratropium, are also administered by inhalation. They antagonize the parasympathetic bronchoconstriction which may be present in some patients with chronic obstructive lung disease.
Methylxanthines, such as theophylline and aminophylline, which are not administered by inhalation, relax bronchial muscle, possibly via phosphodiesterase inhibition.
Bronchodilator therapy is likely to be useful for Carmen as it can open up the airways to some extent and reduce obstruction.

Q8 The dry-powder type of inhaler used to deliver bronchodilators and corticosteroids to the lung by inhalation can sometimes irritate the airways and cause further bronchoconstriction. It is useful to nebulize the drug, that is deliver it to the lung in solution as an aerosol. So the chief advantage of the nebulizer is that it allows drugs to be delivered deep into Carmen's lung, without causing irritation and bronchoconstriction.

Q9 Mucolytic drugs may be useful. These agents facilitate expectoration by reducing the viscosity of sputum. They break bonds in the glycoproteins contained in mucus, so liquefying the secretion and promoting easier removal from the lung.
Anti-inflammatory agents such as the corticosteroids may also be helpful as reduction in airway inflammation reduces obstruction to airflow.

Q10 Pancreatic insufficiency occurs in approximately 80% of cystic fibrosis patients. There is a marked reduction in the water, electrolyte and enzyme content of pancreatic secretion. Because of deficient digestive enzymes, there is inadequate digestion and absorption of nutrients and some nutritional deficiency occurs.
Digestion of carbohydrate begins in the mouth when food mixes with amylase, and protein digestion starts in the stomach when pepsin is released, but the majority of digestion relies on pancreatic enzymes and takes place in the small intestine. In cystic fibrosis the sticky mucus produced blocks ducts in many organs, particularly in the pancreas, and pancreatic secretion is impaired or absent. Diminished digestion and absorption of nutrients leads to malnutrition and slowing of growth in patients with cystic fibrosis.

Most children with cystic fibrosis are diagnosed within one year of birth. Often the symptoms observed in a child, which eventually lead to diagnosis, are malabsorption, failure to gain weight and recurrent respiratory infections.

Q11 Pancreatic enzyme secretion is greatly reduced in cystic fibrosis. Relative or total absence of pancreatic lipase results in failure of fat digestion. The undigested fat is not absorbed, remains in the intestine and is excreted in relatively large quantities. The faeces are pale, bulky, smell unpleasant, float in water and are difficult to flush. This condition is called *steatorrhoea.*

Q12 Because of the deficiency in pancreatic enzymes, emphasis on good nutrition is vital for long-term survival in cystic fibrosis patients. A diet high in calories, fat and protein is required to compensate for losses from malabsorption. Six small meals a day are often better tolerated and more successful in maintaining weight than three larger meals. Enzymes must be added to the ingested food in order to promote the digestion and absorption of what is eaten, and extra salt is usually needed in the diet to replace the heavy salt loss in sweat.

Q13 Fat-soluble vitamins are A, D, E and K.
Cystic fibrosis patients are usually advised to take more than the recommended daily amounts of these vitamins in order to prevent deficiency. A common problem associated with poor absorption of fat-soluble vitamins is deficiency of vitamin K. Vitamin K is required by the liver to produce many blood coagulation factors. Part of the problem for cystic fibrosis patients is their chronic antibiotic therapy, which decreases the bacterial population of the colon: colonic bacteria synthesize vitamin K. Vitamin K deficiency leads to prolonged blood-clotting time. Vitamin D deficiency could cause rickets in a child or osteomalacia in adults. Vitamin A deficiency leads to night blindness, skin and other ocular defects.

Q14 Pancreatic enzyme preparations contain amylase, lipase and protease enzymes. These supplements are given by mouth and compensate for the reduced or absent pancreatic secretions; they assist the digestion of starch, fat and protein. Since the enzymes may be inactivated by gastric acid, they are usually presented in a protected, enteric-coated form which is sprinkled directly on the food.

Key Points

- Patients with cystic fibrosis secrete very viscous mucus in the lung and suffer repeated lung infections. The pancreas is also affected and patients are deficient in pancreatic enzymes; this reduces digestion and absorption of nutrients, so affecting growth.

- The viscous mucus in cystic fibrosis is difficult to clear from the lung: patients need physical therapy and postural drainage to clear the airways.
- Sweat glands have sympathetic cholinergic innervation. Patients with cystic fibrosis secrete a large amount of salt in their sweat and this forms the basis of a diagnostic test for the condition.
- Since cystic fibrosis patients lack digestive enzymes, enzyme preparations containing amylase, lipase and proteases are prescribed in order to improve intestinal absorption of nutrients.

CASE STUDY 32 Chandra's chronic bronchitis

Part 1

Q1 In obstructive disease, tests of expiratory function are the most useful. These include peak expiratory flow measurement, FVC and FEV_1.

In obstructive disease all these values are likely to be reduced. Normal peak flow in a mature man is approximately 500 to 650 l min^{-1}. This may fall to <200 l min^{-1} in obstructive disease. FVC is likely to be 4–5 l in a male adult and may fall to 1–2 l. FEV_1 would be predicted to be 80% of FVC in a normal male, but in obstructive disease it is <70% of FVC.

Q2 *COPD* can be defined as a chronic, slowly progressive disorder characterized by airflow obstruction, which is not fully reversible and does not change significantly over several months. The major forms of COPD are chronic bronchitis and emphysema: both conditions may be present in a patient. Although asthma is also an obstructive disorder, it is usually considered separately. The main difference between asthma and conditions now classified as COPD is the reversibility of bronchoconstriction in the former. In chronic bronchitis and emphysema, the constriction of airways cannot be fully reversed and obstruction progressively increases.

Chronic bronchitis is characterized by increased mucus production and hypertrophy of the mucus glands in the airway mucosa. It is defined as the presence of a chronic or recurrent cough with sputum production on most days, for at least three months of the year, during at least two consecutive years. Patients have hypoxia and retain excess CO_2–they are sometimes referred to as *blue bloaters*.

Emphysema is defined as an abnormal and permanent enlargement of the air spaces distal to the terminal bronchioles with destruction of their walls. Patients with emphysema may have minimal cough and sputum production and retain less CO_2 than bronchitics. They have been called *pink puffers* as they have fast, shallow breathing. Both chronic bronchitis and emphysema result in permanent changes to the structure of the lung and reduction in gas exchange at the respiratory surface.

Q3 Chandra has suffered recurrent chest infections for three years and has had a chronic cough with sputum production during this time. Although he is not a smoker, he has been exposed to occupational dusts in the mining industry, which is known to be associated with development of COPD. His lung function test results are consistent with this diagnosis (see Part 2 of the case study).

Chandra's sputum has changed from grey to green, suggesting a bacterial infection of the chest, and his body temperature is moderately raised: both of these are consistent with an exacerbation of bronchitis. In addition he is cyanosed, his arterial PO_2 is low and he is retaining CO_2. These blood gas abnormalities show that his bronchitis has worsened.

Q4 The expiratory muscles are not used in normal expiration at rest since no muscular effort is required: expiration is a passive process because of recoil of lung structures stretched in inspiration. When there is increased outflow resistance, patients use their thoracic and abdominal muscles to increase intrathoracic pressure and push air from the lung. The muscles used include the internal intercostal muscles of the chest wall and the oblique and transversus muscles of the abdomen.

In patients with chronic obstructive lung disease, air becomes trapped in the lung and total lung volume gradually increases. As the disease progresses, the chest permanently enlarges and the shoulders rise: this shape is often referred to as a *barrel* chest.

Q5 The term *haematocrit* refers to the percentage of total blood volume occupied by packed red blood cells (erythrocytes). In males the haematocrit is normally 40–54%; Chandra's packed cell volume is higher than normal: 59%.

Development of erythrocytes takes place in red bone marrow and is controlled by the hormone erythropoietin, which is produced by kidney cells. The major stimulus for erythropoietin production and release is hypoxia.

Hypoxia can be caused by: (i) decreased ambient oxygen concentration, (ii) hypoventilation, (iii) decreased diffusion across the respiratory surface and (iv) by mismatching of alveolar ventilation with perfusion. In this patient both decreased diffusion and ventilation–perfusion inequality is present. Chandra's blood is low in oxygen: the hypoxia acts as a stimulus for erythropoietin release, which in turn increases red cell production.

Although the raised red cell content of the blood will increase oxygen delivery to tissues, the extra red cells increase blood viscosity, making it more difficult for the heart to pump; blood flow slows and blood pressure increases.

Q6 *Cyanosis* refers to the bluish colour of reduced haemoglobin in the tissues.

When blood flow through tissues is slowed, blood remains in the tissues for a longer time and more oxygen than normal can be extracted. Increased oxygen extraction results in an increased concentration of reduced haemoglobin and makes the skin appear bluish in colour. The blue colour is most easily seen in the lips and mucous membranes.

Chandra's arterial PO_2 is much lower than normal, the percentage of oxygen saturation of his blood is low, and this accounts for the cyanosis: the explanation is supported by the blood test, which shows the reduced arterial PO_2 (72 mmHg) and haemoglobin saturation (81%).

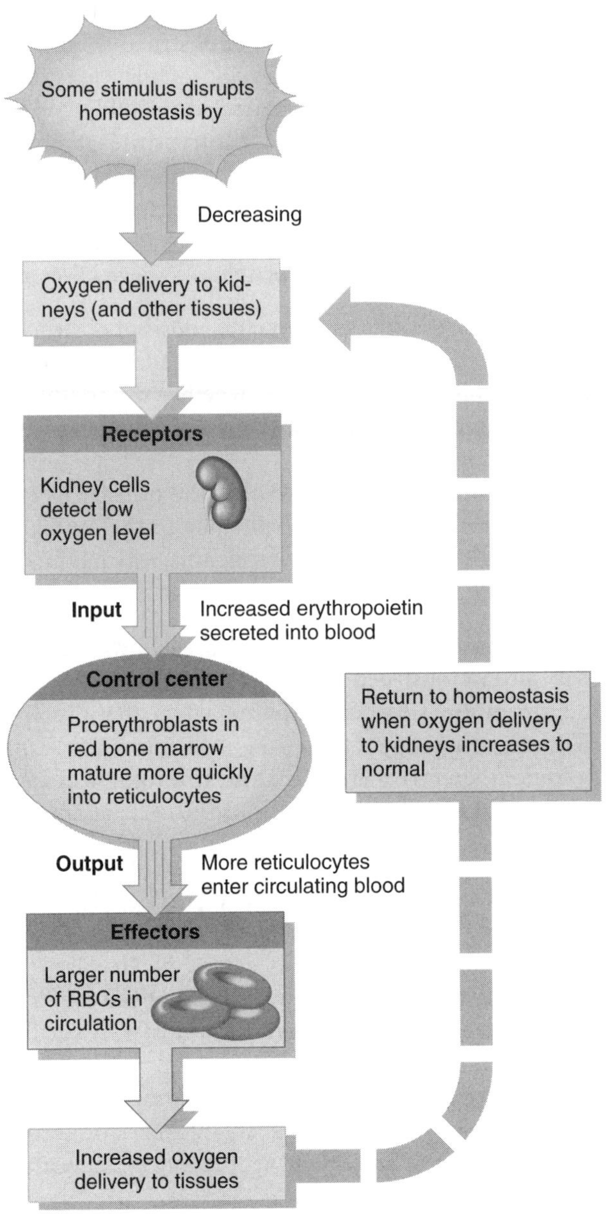

From Tortora and Derrickson *Principles of Anatomy and Physiology, Eleventh Edition 2006.* Reproduced with permission of John Wiley & Sons, Inc.

Part 2

Q7 Chandra and anyone occupationally exposed to chemical fumes and dusts, for example steel workers, farmers and miners or people who are heavy smokers, are at risk of developing COPD. So, chronic bronchitis can be associated with both cigarette smoking and with air pollution. The greater the smoking habit, the higher the incidence of obstructive lung diseases such as bronchitis. However, not all smokers develop COPD. Genetic factors, such as α_1-antitrypsin deficiency, are also risk factors for COPD.

The function of alpha-1-antitrypsin (α_1-antitrypsin) is to oppose the activity of elastases, which are released into lung tissues during the inflammatory process. Lack of this inhibitor allows the elastases to destroy the elastic tissues of the lung, resulting in breakdown of alveolar walls, which accounts for the development of emphysema in some people who have never smoked.

Q8 The majority of the test results support a diagnosis of obstructive pulmonary disease. Chandra's FEV_1 is greatly decreased and the ratio FEV_1/FVC is <0.7, a significant value in determining whether the condition is restrictive or obstructive. A larger residual volume and total lung capacity than normal is typical of obstructive lung disease.

The reduced CO transfer factor shows that the transfer of gas from alveoli to blood is compromised; this is probably due to the ventilation–perfusion inequality usually observed in chronic bronchitis, which limits the respiratory surface area available for gas exchange.

Q9 Bronchoconstriction, airway oedema and breakdown of alveolar walls all contribute to airflow obstruction. Air cannot be easily moved out of the lung during expiration and becomes trapped in the alveoli and small air passages. The air remaining in the lung following a maximal expiration (residual volume) therefore increases. This extra volume of air contributes to an increase in total lung capacity, and over the years alters both the volume and shape of the chest.

Q10 The transfer of CO across the respiratory surface (T_{CO}) can be used to estimate the efficiency of gas transfer in the lung. A small concentration of CO is added to inspired air; it diffuses across the alveolar membranes into the blood. The increase in arterial blood content of CO over a short period of time is measured to estimate the rate of CO transfer. A small concentration of CO must be used as this gas combines strongly with haemoglobin at the same position as oxygen to produce carboxyhaemoglobin.

T_{CO} decreases when alveolar membranes are thickened or fibrosed, when fluid accumulates in the alveoli and when ventilation is uneven or mismatched with perfusion.

Q11 Acute exacerbations of chronic bronchitis can be caused either by viral or bacterial infections. Production of thick, green sputum suggests Chandra has a bacterial infection. Common bacterial pathogens affecting the lung include *Streptococcus pneumoniae* and *Haemophilus influenzae*. It is recommended that COPD patients receive influenza vaccine each year: pneumococcal vaccine is also often recommended in chronic lung disease and may prevent recurrence of chest infection in the elderly.

Q12 Antibiotics are needed by patients with chronic bronchitis as soon as signs of a bacterial infection are present. Chandra has become very breathless at rest, is producing green sputum and has a raised temperature; these are all signs that substantial infection is present. Amoxicillin or erythromycin are usually considered suitable first-line antibiotics for these patients. If the infection is thought to be caused by a viral agent, antibiotics would not be used.

Q13 Normal gas exchange across the respiratory membranes requires that alveoli both receive adequate ventilation and are perfused by blood. If an area of lung has adequate blood flow but reduced ventilation because of airway obstruction, the transfer of oxygen into blood and removal of carbon dioxide will be reduced. As a result arterial PO_2 falls and arterial PCO_2 rises (hypercapnia). In a normal lung, low arterial PO_2 stimulates constriction of the pulmonary blood vessels, diverting blood away from hypoxic areas to areas of lung that have better ventilation. In obstructive lung disease the reduced ventilation is widespread, few areas receive an effective air supply and the vasoconstrictor mechanism does not benefit gas exchange. Hypoxia and hypercapnia then persist.

Q14 Short-acting bronchodilators such as the beta-2-agonist (β_2-agonist) salbutamol reduce airflow limitation and have been shown to benefit bronchitic patients. Their effectiveness should be assessed in each patient, using tests of expiratory function both before and following inhalation of the agent. It may be necessary to use higher doses of β_2-agonists in COPD than are used in asthma. Parasympathetic-induced bronchoconstriction may also be present in COPD, and some reversal of bronchoconstriction can be achieved using a muscarinic antagonist such as ipratropium or oxitropium. The muscarinic antagonists show a slower bronchodilator effect than the β_2-agonists.
Recent clinical guidelines suggest that a trial of a corticosteroid inhaler may be useful and should be made in bronchitic patients. Not all patients will benefit, but if the trial shows steroids to be effective they can be added to the patient's medication as maintenance therapy.

Q15

(1) Pure O_2
Administration of oxygen-rich gas mixtures is useful in hypoxia, but 100% O_2 is not often used. In chronic bronchitis, hypoxia and hypercapnia coexist, the respiratory centre in the medulla becomes tolerant to the high CO_2 content of blood and is relatively insensitive to it. Respiratory drive is maintained by hypoxia acting via chemoreceptors in the aorta and carotid body. Removal of the hypoxic stimulus to the respiratory centre in the medulla may actually stop the patient breathing.
In COPD a lower concentration of oxygen, 24–28%, is usually used. The aim is to increase the arterial PO_2, without worsening CO_2 retention and respiratory acidosis.

(2) Muscarinic agonist
Parasympathetic stimulation increases bronchoconstriction via muscarinic cholinoceptors. A muscarinic agonist will increase bronchoconstriction, so this drug is definitely not recommended.

(3) Cromoglicate
Sodium cromoglicate is used for prophylaxis in asthma when there appears to be an allergic basis to the condition. Although its mechanism of action is not well understood, it appears to reduce the release of inflammatory agents from mast cells and so is useful in asthma prophylaxis, particularly in children. It is unlikely to be of value in COPD.

(4) Aspirin
Aspirin is a non-steroidal anti-inflammatory drug (NSAID) which is used to treat pain and, in low dosage, for the prophylaxis of coronary heart disease. It must be used with caution in asthmatic patients because of possible bronchoconstriction, and in large doses aspirin can adversely affect respiration by depressing the respiratory centre, leading to CO_2 retention. Therefore it is not useful in COPD.

Key Points

- Chronic bronchitis is an obstructive pulmonary disease linked to smoking and prolonged working in dusty environments and is characterized by excessive mucus production with repeated chest infections.
- FVC and FEV_1 are reduced and the FEV_1/FVC ratio is < 0.7.
- Blood gas tensions are usually abnormal: arterial PO_2 is lower and arterial PCO_2 higher than normal. Patients may be cyanosed because of an increase in reduced (deoxygenated) haemoglobin in the tissues.
- Patients may have an increased haematocrit (percentage of blood occupied by erythrocytes) since erythropoiesis is stimulated by the ongoing hypoxia associated with bronchitis.
- The efficiency of oxygen diffusion across the respiratory membrane can be estimated by performing a test of T_{CO} transfer capacity.
- Common bacteria causing exacerbation of chronic bronchitis include *S. pneumoniae* and *H. influenzae*. The antibiotics usually prescribed for chest infections in patients with chronic bronchitis are amoxicillin and erythromycin.

6

Kidney and body fluid disorders

CASE STUDY 33 Greg's glomerulonephritis

Part 1

Q1 The glomerulus is a ball of capillaries which is part of the renal corpuscle; the other portion of this structure is Bowman's capsule, which forms the start of the nephron. The wall of Bowman's capsule is composed of a layer of specialized epithelial cells with extensions or foot processes which are in contact with the glomerulus and are called *podocytes*. The gaps between the foot processes are known as *slit pores*. These pores allow small molecules to pass through the epithelial layer into the nephron tubules. Below the epithelium is a basement membrane which prevents the passage of large proteins and whole cells into the renal tubules.

Blood enters the glomerulus in the afferent arteriole. As it passes through the glomerular capillaries, fluid filters across the capillary wall into the renal tubules. Blood leaves the glomerulus in the efferent arteriole, which then gives rise to peritubular capillaries surrounding the renal tubules, and the vasa rectae which follow the loops of Henle down into the medulla.

Q2 Water and small molecular weight (MW) substances such as glucose and amino acids dissolved in plasma are filtered from the blood and enter the nephron at the same concentration as blood plasma. All blood cells and large

Clinical Physiology and Pharmacology Farideh Javid and Janice McCurrie

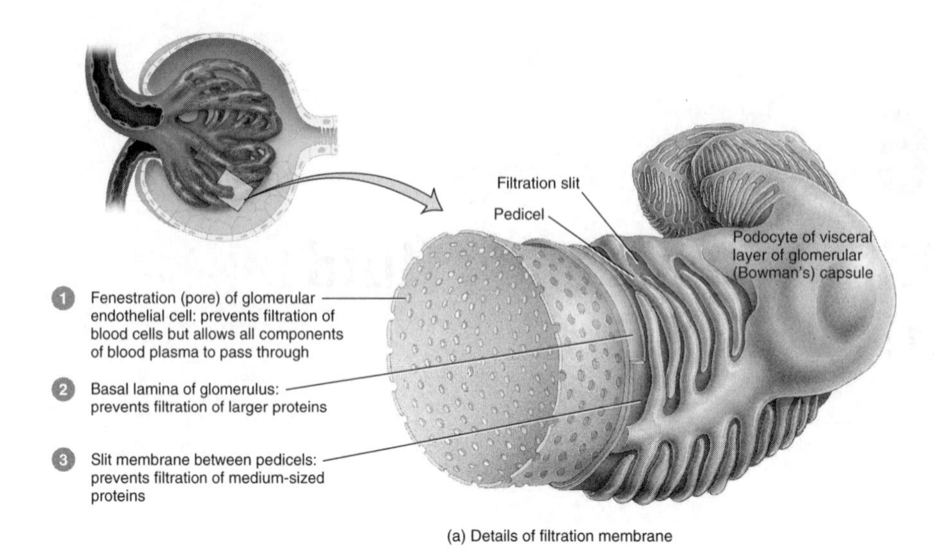

Filtration membrane. From Tortora and Derrickson *Principles of Anatomy and Physiology, Eleventh Edition 2006*. Reproduced with permission of John Wiley & Sons, Inc.

MW substances, including plasma proteins and the drugs and hormones bound to these proteins, are unable to cross the normal glomerular membrane, so they are retained in the blood. However, tiny amounts of small proteins such as albumin are able to squeeze through the membrane, and normal urine often contains traces of albumin.

In inflammatory diseases and infections of the glomerulus, the filtration membrane becomes leaky and considerable amounts of plasma protein are lost from the blood into the renal tubules. The appearance of protein and blood in the urine suggests there is a problem with the glomerular membrane.

Q3 Tissue fluid is derived from blood plasma in the capillaries. Capillaries have thin walls, one cell thick. Their function is to exchange nutrients and waste materials between blood and tissues. The capillary is permeable; most small solutes move across the wall by simple diffusion. The forces that move fluid across the capillary membrane are hydrostatic and osmotic. Hydrostatic pressure provided by the heart tends to push fluid, without protein, out of the arterial end of the capillary into the tissues, and this fluid loss increases the osmotic pressure of blood remaining in the capillary. The osmotic effect of plasma proteins remaining in the capillary blood as it flows towards the veins favours the influx of fluid from the tissues back into the blood. These hydrostatic and osmotic forces are sometimes called *Starling forces*, after the physiologist who first described the process.

There is normally a balance between the volume of fluid leaving the capillary and that flowing back from the tissues at the venous end of the capillary. Any excess fluid remaining in the tissues is returned to blood via the lymphatic system.

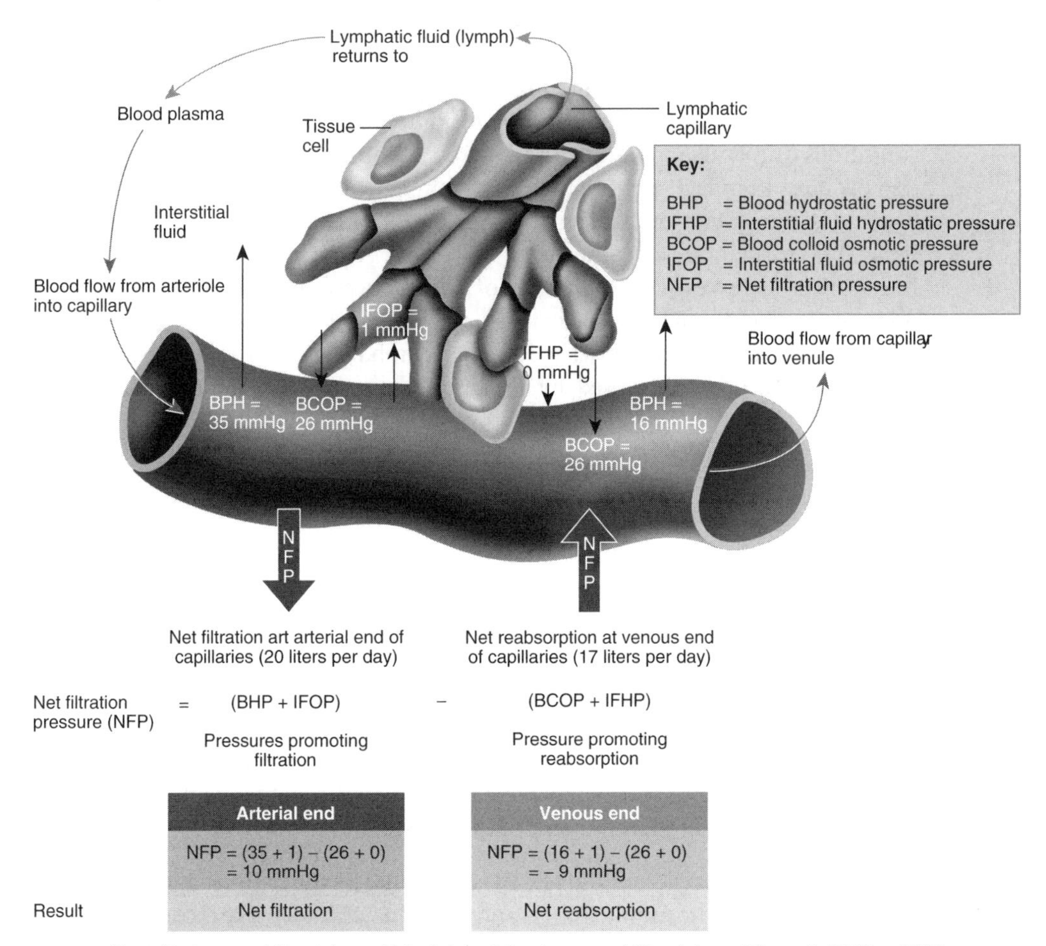

From Tortora and Derrickson *Principles of Anatomy and Physiology, Eleventh Edition 2006.* Reproduced with permission of John Wiley & Sons, Inc.

Q4 If protein is lost from the body, for example because of kidney disease, or there is a reduction in the synthesis of plasma protein, for example in starvation, the balance of fluid loss and gain in the capillaries is altered. Reduction in plasma protein reduces the oncotic pressure and reduces the return of fluid from the tissues back to the capillaries. So fluid accumulates in the tissues and forms oedema.

When oedema fluid collects in the tissues of the skin, it gives a puffy look to the skin of the face. In the lung, the capillaries run close to the alveoli, and reduction in plasma oncotic pressure can result in fluid accumulation in the alveolar wall and in the alveoli. This fluid increases the diffusion distance for oxygen between blood and alveolar air and acts as a diffusion barrier, reducing gas exchange. If severe, lung (pulmonary) oedema can result in development of abnormal blood gas concentrations. Treatment of pulmonary oedema is critical as it can develop into a life-threatening situation.

Q5 Many factors can cause oedema. Anything which reduces plasma protein can cause oedema, for example starvation, liver disease (liver produces plasma proteins) and burns (because plasma and its protein is lost from the surface of burned skin). Increased capillary permeability in allergic or inflammatory conditions causes oedema. High arterial pressure or high venous pressure can also produce it; in fact, anything which disturbs the hydrostatic and osmotic balance in the system or changes the capillary permeability which regulates tissue fluid formation can cause oedema.

In addition an excessive production of some hormones, for example the mineralocorticoids such as aldosterone, can cause salt and water retention, which results in oedema. Blockage of the lymphatic system or damage to lymph vessels, perhaps caused by radiation therapy, can produce a local oedema.

Part 2

Q6 *Dyspnoea* is a general term which refers to breathing problems which patients may describe, for example breathlessness or shortness of breath. Greg has developed pulmonary oedema, which can reduce the diffusion of gases across the alveolar membrane, causing a decrease in arterial PO_2 together with increase of arterial PCO_2 and symptoms of breathlessness.

Oedema develops in tissues such as the lung in kidney disease; it is mainly due to the large loss of albumin in the urine. Albumin loss reduces the oncotic pressure of plasma and so disrupts the normal formation of tissue fluid from blood plasma, leading to movement of extra fluid into the alveoli.

Q7 The volume of extracellular fluid (ECF) is regulated by the renin–angiotensin system, antidiuretic hormone (ADH) and the kidney. Fluid intake is controlled by altering drinking behaviour, and thirst is regulated by a centre in the hypothalamus. Water is also produced in the body as a result of oxidative metabolism. Water remains in the ECF only if accompanied by an osmotic equivalent of sodium, which is the main extracellular cation. When fluid volume and sodium content of the body is low, the renin–angiotensin system is stimulated. Renin is an enzyme produced in the kidney which acts on the protein angiotensinogen. The resulting angiotensin stimulates the release of the salt-retaining hormone aldosterone from the adrenal cortex. As a consequence, salt and water are retained by the kidney to increase blood volume. Also, when ECF volume decreases, volume receptors in the atria are activated and ADH is released from the posterior pituitary gland to favour water absorption by the kidney.

Q8 The antibodies produced in the body in response to a streptococcal infection combine with bacterial antigens to form complexes, which become trapped in the glomerular capillaries. Inflammatory changes are produced in the glomerular filtration membrane, which alter its permeability. The inflamed glomerular membrane becomes very leaky, allowing proteins and blood cells, which normally cannot pass into glomerular filtrate, to enter the proximal tubule and be excreted in the urine.

Q9 Loss of albumin through the glomerular membrane reduces the concentration of albumin in the blood. Albumin plays a major role in the maintenance of ECF volume, and when there is a deficiency additional fluid passes from plasma into the tissues to form oedema. Passage of extra fluid from the circulation into the tissues reduces blood volume, which stimulates the renin–angiotensin system and also triggers the thirst mechanism via osmoreceptors in the hypothalamus.

Q10 The condition is treated with antibiotics to eliminate remaining bacteria, salt and water restriction to limit oedema and a diuretic such as furosemide to mobilize existing pulmonary oedema.

Q11 Simple concentration and dilution tests can be used to check whether the regulatory mechanisms of the kidney are operating normally. There are also simple dipstick tests for the presence of protein and other abnormal constituents in urine: the absence of these abnormal constituents is an indication that renal function has returned to normal.

As renal function improves, the excretion of urea increases and the concentration of urea in blood declines. So a reduction in blood urea nitrogen (BUN) is also a useful sign of returning kidney function. More complex tests, such as creatinine clearance, would be needed to check whether the glomerular filtration rate (GFR) has returned to normal.

Key Points

- The glomerulus is a ball of capillaries situated in Bowman's capsule, which is the first part of the nephron. Bowman's capsule is composed of specialized cells, podocytes, which, with the capillary walls, form a filtration membrane. Water, ions and substances of low MW can pass through the filtration membrane into the nephron, leaving proteins and cells behind in the blood.
- In inflammatory conditions and infection, the filtration membrane becomes leaky and large amounts of protein, particularly albumin, may escape into the nephron and appear in the urine.
- Tissue fluid is formed at the arterial end of the capillaries and carries oxygen and nutrients through the tissues. It is reabsorbed at the venous end of the capillaries caused by osmotic effects of plasma proteins. Excess fluid in the tissues is normally removed by the lymphatic system.
- Accumulation of fluid in tissues causes oedema, which can make facial skin look puffy and cause fluid accumulation in the alveoli (pulmonary oedema), reducing gas exchange. Oedema is observed in kidney disease, when large quantities of albumin are lost from blood and excreted in urine.
- Glomeruli can be damaged as a result of an apparently minor streptococcal throat infection. Alterations in renal function and return to normal function can be estimated by renal function tests such as concentration/ dilution tests and estimation of GFR.

CASE STUDY 34 Kevin's chronic kidney problems

Part 1

Q1 The capillaries in the glomerulus have thin, permeable walls, one cell thick. The capillaries are of the fenestrated type, and the endothelial cells are in direct contact with a basement membrane. In turn the basement membrane is in contact with podocytes, specialized cells of Bowman's capsule. The filtration membrane of the nephron therefore consists of the fenestrated endothelial cells, the basement membrane and the podocyte membrane. The ultrafiltration of blood plasma which occurs through this thin membrane separates blood cells and proteins from the fluid components of plasma. The pore size of the filtration membrane is roughly the diameter of albumin, the most plentiful protein in blood. However, normally only a very small quantity of albumin slips through the filtration membrane as the basement membrane has a net negative charge, which repels the negatively charged albumin. The net filtration pressure is approximately 10 mmHg, and GFR is normally approximately 125–130 ml min^{-1}. One hundred and eighty litres of fluid are usually filtered into the renal tubules each day. The majority of this fluid, around 99%, is reabsorbed by the renal tubules since the urine volume of an adult is normally only about 1–2 l per day.

Q2 Type 1 diabetes mellitus that is poorly controlled is associated with damage to renal blood vessels and changes in the glomerular membrane. The membrane can become very leaky and may allow large amounts of protein into the urine. Since the majority of protein in the plasma is albumin and albumin is the smallest of the plasma proteins, it passes easily through the damaged membrane in significant amounts.

The main function of albumin in the plasma is to provide colloid osmotic pressure. It is of major importance in maintaining blood volume and in the exchange of fluid between blood and the tissues. Heavy proteinuria may involve the loss of >3.5 g of albumin per day and this, in turn, causes a reduction in plasma oncotic pressure. When plasma oncotic pressure is reduced, fluid is not completely reabsorbed from the tissues at the venous end of capillaries. The fluid is retained within the tissues, causing oedema. The effects of gravity on fluid accumulation in the body causes oedema to be more marked in the lower body than in the upper parts, so oedema is often noticed first around the ankles.

Q3 Creatinine is a product of muscle metabolism. It is released from muscle into the blood at a fairly constant rate and is normally excreted in the urine at the same rate so that plasma creatinine concentration is constant. Creatinine is filtered at the glomerulus and is not reabsorbed; it is excreted by the renal tubules unchanged. Creatinine excretion can therefore give us an estimate of the rate of filtration in the glomerulus, and creatinine clearance is used to estimate GFR. If plasma creatinine rises, it suggests that the filtration rate has diminished; this is a sign of disturbed renal function. Measurement of GFR can be made by measuring the creatinine concentration in blood and urine using the formula:

$$C = \frac{UV}{P}$$

where C = clearance (ml min^{-1}), U = concentration in the urine (mg ml^{-1}), V = urine volume (ml min^{-1}) and P = plasma concentration (mg ml^{-1}).

Q4 Potassium concentration is mainly controlled by the steroid hormone aldosterone. Aldosterone release from the adrenal cortex can be stimulated by either decreased plasma sodium or by increased plasma potassium concentration. An increase in aldosterone secretion causes retention (reabsorption) of sodium in the distal nephron in exchange for secretion of potassium into the urine. The amount of potassium excreted by the kidney is influenced by the acid–base status of the body. In alkalosis, potassium excretion increases, whereas in acidosis it is decreased. In the distal nephron H^+ and K^+ compete for excretion in exchange for the reabsorption of sodium. Insulin also affects plasma potassium concentration because it promotes the movement of potassium from the plasma into cells.

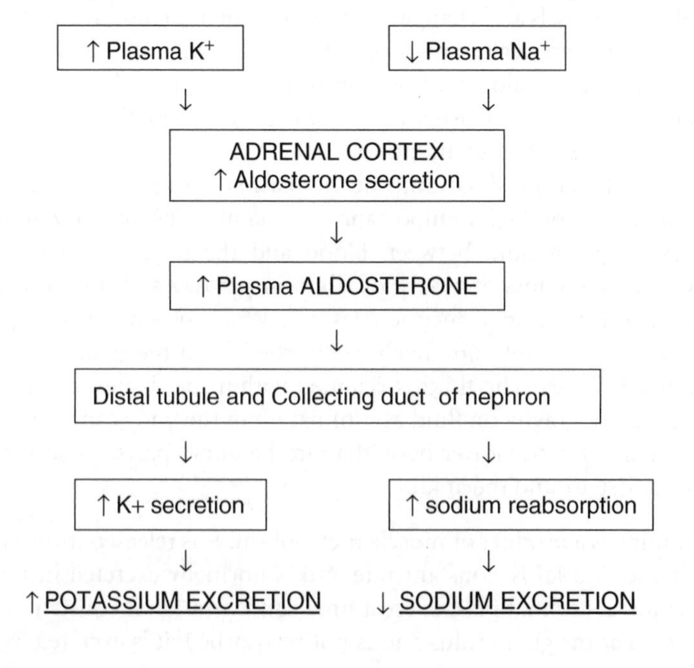

Q5 Diabetes is one of the leading causes of renal failure, and ureamia is a syndrome of renal failure. The term *uraemia* covers both increased blood concentration of urea and increased creatinine concentration. Uraemia affects all tissues of the body and can cause symptoms such as anorexia, nausea, vomiting, fatigue, drowsiness, headache and neurological changes. So a high blood urea concentration, which is associated with chronic renal failure, can account for many of Kevin's symptoms.

Part 2

Q6 Most of the potassium in the body is located inside cells; only 2% of total body potassium is found in the ECF. However, the regulation of potassium in the ECF is particularly important for the function of all excitable tissues. The excitability of nerve and muscle depends on their resting membrane potential, which in turn depends on the concentration gradient for K^+ across the plasma membrane. Hyperkalaemia causes muscle weakness and may lead to paralysis. Disorders of cardiac rhythm are also likely to occur and may progress to cardiac arrest. The earliest signs of hyperkalaemia can be seen in the electrocardiogram (ECG) as peaking, or 'tenting', of the T waves (T waves represent repolarization of the ventricles), followed by a widening of the QRS complex.

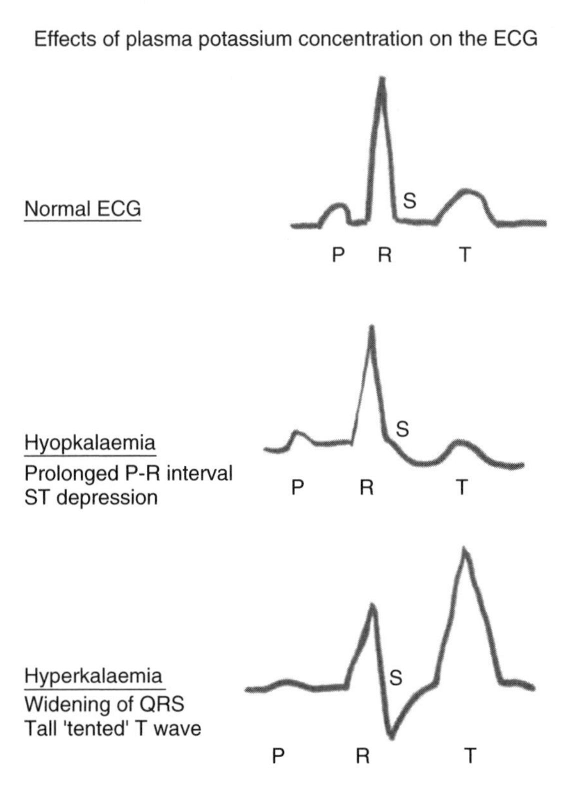

Kevin's [K^+] is somewhat higher than the normal range and a high plasma potassium concentration could account for his weakness and possibly contributes to his nausea. However, the increased plasma urea concentration, which Kevin also shows, is known to cause nausea and vomiting and is more likely to be responsible for these symptoms.

Q7 Calcium is present in both intracellular fluid (ICF) and ECF, but the concentration in the ECF is twice as high as that in the ICF. Calcium is found in both ionized and bound forms, and Ca^{2+} homeostasis is mainly controlled by parathyroid hormone, which increases absorption of calcium in the intestine and reabsorption in the nephron. Calcitonin also affects ECF calcium concentration by promoting renal excretion when there is an excess of calcium in the body. The normal kidney filters and reabsorbs most of the filtered calcium; however, in renal disease this is reduced and blood calcium decreases.

Calcium and phosphate imbalance can occur in patients with renal failure, leading to osteomalacia (defective mineralization of bone). Osteomalacia is mainly due to reduced production of 1,25-dihydroxycholecalciferol, an active form of vitamin D metabolized in the kidney. Deficiency of 1,25-dihydroxycholecalciferol reduces the absorption of calcium salts by the intestine.

Q8 The thiazide diuretics reduce urinary excretion of calcium and can also be used to decrease the likelihood of calcium-based renal stones. On the other hand, the loop diuretic furosemide reduces reabsorption of calcium and increases calcium excretion.

Q9 The signs and symptoms of anaemia may include: tiredness, headache, dizziness, fainting and breathlessness. Patients with more serious anaemia may suffer from tachycardia, palpitations and, if anaemia is severe, angina (pain in the chest) during exercise.

Q10 Adult red blood cells are produced by the bone marrow at the ends of long bones and in the pelvis, skull, ribs and sternum. In response to severe anaemia the active bone marrow in the long bones becomes more extensive. Normally, the total number of circulating red blood cells is maintained constant. Production is stimulated by the glycoprotein erythropoietin (EP), which is mainly produced by the endothelial cells of the kidney. EP production is stimulated by hypoxia and a decrease in haemoglobin concentration. EP stimulates the stem cells in bone marrow to differentiate into mature erythrocytes.

The anaemia of chronic renal disease is due to the reduced secretion of EP by the kidney, and in renal failure red cell count and haemoglobin concentration fall considerably.

Key Points

- Chronic renal failure may develop in patients with poorly controlled diabetes mellitus and is characterized by heavy proteinuria, weakness, tiredness, nausea and abnormal concentrations of creatinine, electrolytes and urea (uraemia) in blood. There may also be symptoms of headache and neurological changes.
- Loss of albumin in the urine following glomerular dysfunction causes oedema, which is often first seen in the dependent parts of the body, for example ankles.
- Potassium (K^+) and calcium (Ca^{2+}) levels in blood are affected by renal failure. High K^+ leads to muscle weakness and may cause cardiac rhythm disturbance. The low blood Ca^{2+} leads to defective mineralization of bones (osteomalacia)
- Patients in renal failure may become anaemic because of deficient renal secretion of EP. EP stimulates production of red cells in bone marrow.

CASE STUDY 35 The polar bear's fun run

Part 1

Q1 The sensation of thirst is produced by a decrease in ECF volume and by an increased plasma crystalloid osmotic pressure. The sensation is usually elicited when body water loss is about 2% of an individual's body weight. The changes

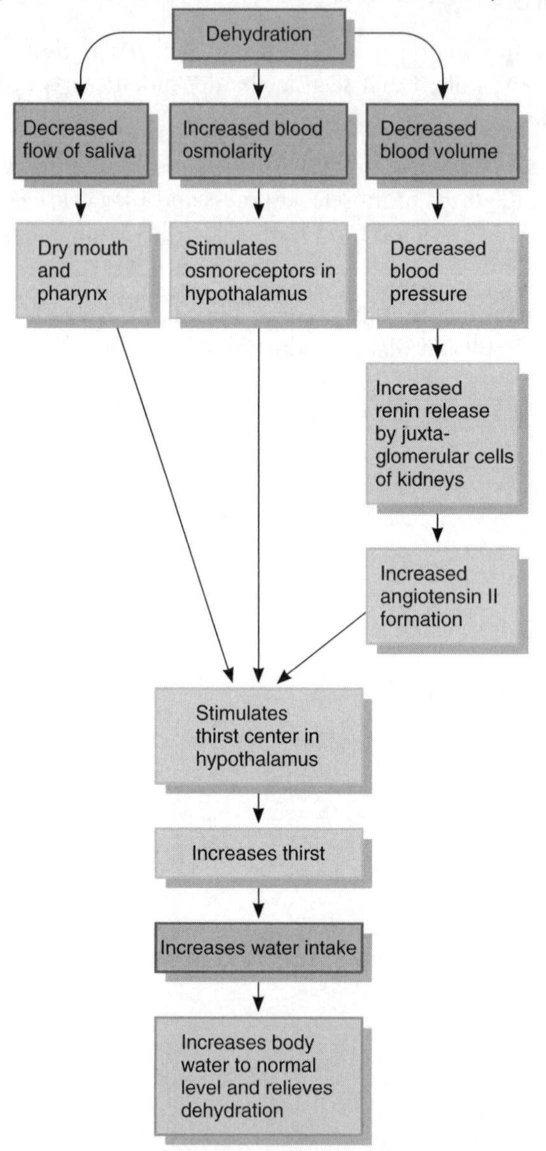

From Tortora and Derrickson *Principles of Anatomy and Physiology, Eleventh Edition 2006.* Reproduced with permission of John Wiley & Sons, Inc.

in ECF volume also stimulate ADH production, which in turn stimulates the kidney to retain water. The brain region associated with the sensation of thirst is located in the hypothalamus; the receptors, osmoreceptors, are located close to the site of ADH production. In addition, angiotensin released when blood volume or pressure, or sodium concentration of body fluids, decreases also elicits the sensation of thirst. Dryness of the mucous membranes of the mouth and pharynx make one thirsty; as a person drinks, these membranes are moistened and the sensation of thirst diminishes. Drinking water restores the depleted ECF and plasma volume but dilutes ECF osmolarity. As ECF volume and osmolarity become normal, the thirst sensation declines further and drinking stops.

Q2 Sweat is a hypotonic solution containing water, sodium and chloride. Loss of sweat decreases ECF volume and increases its osmolarity. If the volume lost is not great, the kidney can compensate by retaining extra sodium and water from the glomerular filtrate. But when loss of fluid via sweat is severe, the compensatory mechanisms may cause the kidney to stop producing urine for a time (anuria).

Salt and water balance are closely related. Water can remain in the ECF only if accompanied by sodium ions, which are the major cations in the ECF and form 90% of the total cation content. If water is added to the plasma without an appropriate amount of sodium ions to maintain normal osmotic pressure, the water will leave the ECF and move into the body cells. Although thirst is known to be a powerful stimulus to drink and replace the lost water, the corresponding stimulus for salt intake or salt 'appetite' is poorly understood and is probably of minor importance in human subjects.

Q3 The sodium content of the body is regulated by two hormones, aldosterone and atrial natriuretic peptide. The kidney maintains the normal ECF sodium concentration in a narrow range around 142 $mEq\,l^{-1}$, by adjusting the renal tubular absorption of sodium. Fine control of sodium involves aldosterone, which increases sodium reabsorption in the distal part of the nephron. The major stimuli for aldosterone release from the adrenal cortex are a fall in sodium concentration of the plasma, a decrease in blood pressure (BP) or a rise in plasma potassium concentration. These changes act via secretion of renin from the juxtaglomerular cells of the kidney. Renin stimulates angiotensin formation from the large protein angiotensinogen. Angiotensin is a potent vasoconstrictor which acts on the adrenal cortex to release aldosterone.

The natriuretic peptides are produced and released from various tissues, including the atria, and cause the excretion of additional sodium and water by the kidney, so reducing body sodium content, blood volume and BP.

Q4 Dehydration reduces the volume of ECF, a compartment which includes blood volume. Blood volume must be maintained constant, even at the

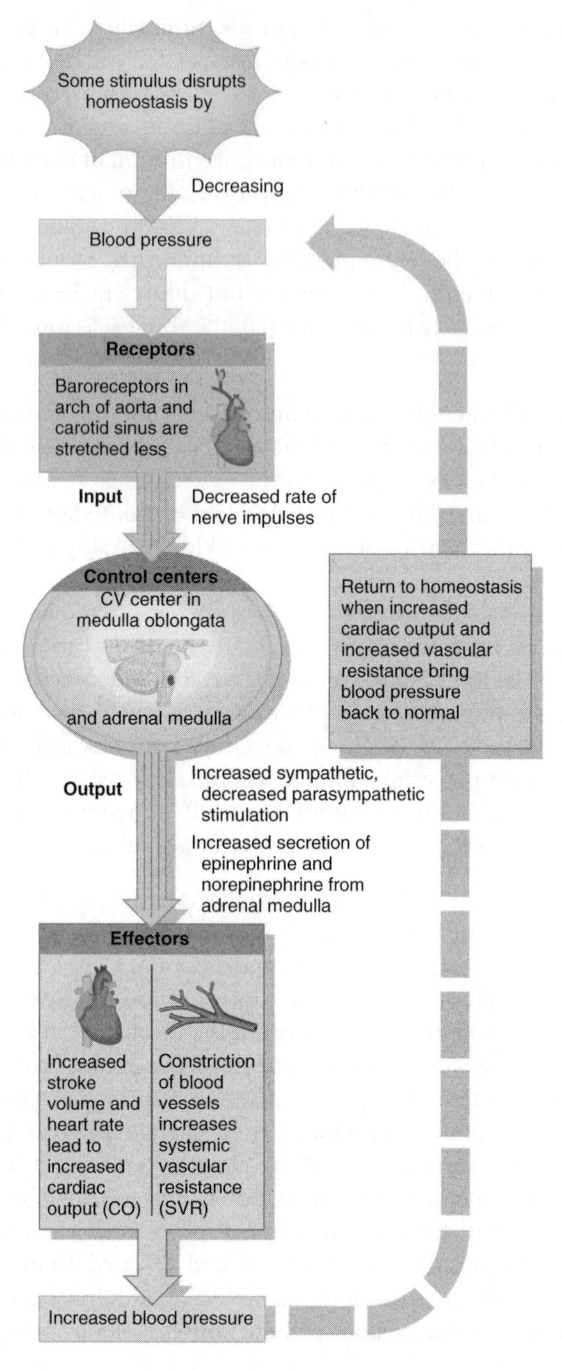
Some stimulus disrupts homeostasis by
Decreasing
Blood pressure
Receptors
Baroreceptors in arch of aorta and carotid sinus are stretched less
Input
Decreased rate of nerve impulses
Control centers
CV center in medulla oblongata
and adrenal medulla
Return to homeostasis when increased cardiac output and increased vascular resistance bring blood pressure back to normal
Output
Increased sympathetic, decreased parasympathetic stimulation
Increased secretion of epinephrine and norepinephrine from adrenal medulla
Effectors
Increased stroke volume and heart rate lead to increased cardiac output (CO)
Constriction of blood vessels increases systemic vascular resistance (SVR)
Increased blood pressure

expense of a reduction in intracellular water, in order to keep the circulation operating normally. Initially, some compensation for reduced ECF volume occurs because of movement of water from the body cells into the blood, but when the dehydration continues and becomes more severe the compensation is inadequate, blood volume falls and venous return is reduced. Diminished venous return reduces stroke volume and cardiac output and so, in turn, decreases BP. The fall in BP elicits a reflex via the baroreceptors, which tends to increase BP towards normal.

Q5 The baroreceptor reflex, which follows the fall in BP caused by reduced ECF volume (hypovolaemia), causes vasoconstriction and increases the heart rate in an effort to raise BP back towards normal. So Eddie's racing pulse is the result of a sympathetic reflex. But the reduction in BP is likely to make Eddie feel dizzy, because of reduced perfusion of the brain, and in this situation fainting (syncope) often occurs. Fainting is actually beneficial because, when someone falls to the ground, the effects of gravity on the circulation are minimised and a person's BP improves.

Q6 In heavy sweating over 1 l of water may be lost per hour. The loss of large amounts of salt and water during prolonged, vigorous exercise in a very hot environment can cause a substantial decrease in the volume of the ECF compartment and produces a powerful sensation of thirst. The intake of a large volume of water helps to expand the ECF quickly but in replacing the lost water the ECF will be diluted and the concentration of sodium reduced, causing a decrease in crystalloid osmotic pressure. As the osmotic pressure of the ECF falls, water will move from this compartment into the body cells, causing cellular swelling. So any water taken into the body gradually becomes distributed in, and dilutes, both the extracellular and intracellular compartments.

Dilution of the ECF reduces the thirst sensation and reduces ADH production so that normally the kidney produces diluted urine and eliminates the extra water. However, following rapid dehydration, the reduction in blood volume decreases cardiac output and BP, causing reflex vasoconstriction and reduced perfusion of tissues, including the kidney. This in turn reduces GFR and urine formation (oliguria) and the water is retained in the body.

Reduction in sodium content and osmotic pressure of body fluids affects muscle and nerve performance, causing muscular weakness. Although most tissues are actually fairly tolerant of moderate cellular swelling, the cells in the brain lie in a space restricted by the rigid bones of the skull. As the cells swell, cerebral pressure builds up and neurological changes occur, leading to disorientation and confusion.

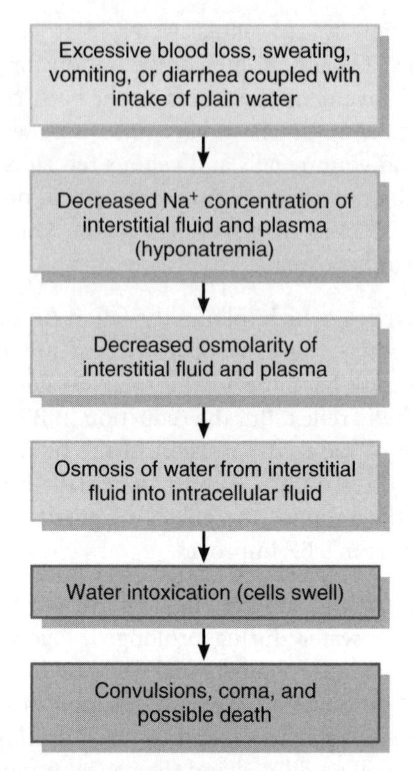

From Tortora and Derrickson *Principles of Anatomy and Physiology, Eleventh Edition 2006.* Reproduced with permission of John Wiley & Sons, Inc.

Part 2

Q7 Replacing the fluid lost in sweat by drinking a large volume of water and receiving an IV infusion of dextrose solution would contribute to a significant dilution of body fluids, and the test result shows that plasma sodium concentration is reduced below the normal value; this condition is termed *hyponatraemia*. The haematocrit is a measure of the percentage of blood volume contributed by the red blood cells. Since the volume of all ECF compartments, including blood volume, is increased by the water load, the number of red blood cells per millilitre of blood is reduced; this is shown by the reduction in haematocrit from the normal male range of approximately 41–54%. In severe cases the condition of hyponatraemia can lead to pulmonary and cerebral oedema.

Q8 Dehydration reduces ECF volume, venous return, cardiac output and BP, leading to both a reflex vasoconstrictor response and to the stimulation of the renin–angiotensin system. Angiotensin stimulates the release of aldosterone from the adrenal cortex, which causes salt and water retention in the distal tubule and collecting ducts of the nephron. The same stimuli release ADH so that more water is conserved by the kidney. Together the aldosterone and ADH mechanisms can compensate for the water loss, unless dehydration is very severe. In the latter case the circulation may collapse unless a suitable volume of isotonic fluid is supplied.

Q9 ADH, or vasopressin, is produced in the supra-optic nucleus of the hypothalamus. Stimulation of sensory receptors in the circulation which respond to stretch, for example in the atria, and of osmoreceptors in the hypothalamus which results in secretion of ADH from hypothalamic neurones. This region of the hypothalamus is connected to the posterior pituitary gland. ADH is enclosed in small vesicles and moves down axons to accumulate in the posterior pituitary gland, from which it is released when blood volume decreases or plasma osmotic pressure increases. ADH regulates the permeability of the renal collecting ducts to water. In the presence of a high circulating ADH concentration, this area of the kidney becomes very permeable and water reabsorption is maximal so that a small volume of concentrated urine is produced. In the absence of ADH, the water permeability of the collecting ducts is low and a large volume of dilute urine is produced. ADH has no effect on sodium or chloride reabsorption.

Q10 An excessive amount of ADH may be produced by some brain tumours, certain drugs and some lung cancers. Since this causes a reduced water excretion, an excess of ADH leads to water retention and hyponatraemia. As hyponatraemia in the ECF causes passage of water into the body cells, there may be brain swelling, raised intracranial pressure and neurological symptoms, such as headache, muscle weakness, lethargy, nausea and vomiting, irritability, confusion and coma.

Damage to the hypothalamus may lead to reduced ADH secretion. This causes diabetes insipidus: there are large losses of water in the urine (polyuria) and up to 5–10 l of urine may be produced in a day. Symptoms include dehydration, weight loss, polydipsia (excessive drinking) and polyuria. In some patients ADH is produced but the kidney is insensitive to it, this is termed *nephrogenic diabetes insipidus*. The symptoms are similar to those described above: patients are unable to concentrate urine and suffer from constant thirst.

Key Points

- Thirst is sensed in the hypothalamus and occurs when extracellular volume decreases or osmotic pressure increases. ADH is then released from the posterior pituitary, increases water reabsorption in the nephron and reduces the volume of urine produced.

- Body sodium content is primarily regulated by aldosterone, which stimulates renal reabsorption of sodium. Atrial natriuretic peptide decreases sodium reabsorption by the nephron and increases its excretion in urine.
- Large amounts of salt and water can be lost in sweat, which may cause loss of 1 l of fluid per hour. Severe loss of fluids by this route may stop urine production.
- Rapid replacement of fluid loss by drinking water dilutes the ECF, reducing the sodium concentration (hyponatraemia) and haematocrit, which causes a shift of water into the cells. Cellular swelling causes detrimental effects on excitable cells, with symptoms of muscle weakness and disorientation.

CASE STUDY 36 The housewife who drank too much

Q1 *Polydipsia* refers to excessive drinking. *Polyuria* means an excessive excretion of urine.

Q2 Water is produced as a result of metabolism, approximately 200 ml per day, water in food accounts for approximately 700 ml per day and the average daily intake of liquids is approximately 1.5 l.
In addition to the excretion in urine, water is lost from the body in several ways.
Water loss via the skin is approximately 350 ml per day, a similar volume is lost from the lung in expired air. One hundred millilitres per day is lost in normal faeces and a similar volume may be excreted in sweat.
The volume of urine produced each day is very variable since the kidney maintains water balance in the body by adjusting the excretion of water, keeping the osmotic pressure of body fluids constant. Approximately 1.5 to 2 l of urine are usually excreted each day.

Q3 Vasopressin is a small peptide hormone consisting of nine amino acids, most of which is synthesized in neurosecretory cells of the supraoptic nucleus of the hypothalamus. Small quantities are also produced in the neighbouring paraventricular nucleus. The hormone is transported down the axons of the neurosecretory cells via the infundibulum to the posterior pituitary, where it is stored until release into the blood is triggered by nerve impulses from the hypothalamus. Vasopressin is better known as *antidiuretic hormone* (ADH). The name vasopressin relates to its vasoconstrictor action, which increases pressure in the vascular system. This action was discovered before its effects on water retention were known.
Release of vasopressin occurs when:

- hypothalamic osmoreceptors detect an increase in the osmotic pressure of the ECF
- circulating blood volume, detected by cardiovascular volume receptors, decreases
- arterial BP, detected by baroreceptors, decreases
- angiotensin concentration increases.

Q4 An important process in concentrating urine is the creation of an osmotic gradient in the renal medulla. This is produced by active pumping of salts out of the ascending limb of the loop of Henle, without accompanying water, since the ascending limb is impermeable to water. The pump involves a $Na^+/K^+/2Cl^-$ coupled co-transporter located in the ascending limb cells, which moves the ions into the interstitial fluid of the medulla. The small osmotic gradient produced by this process is multiplied by the counter-current flow of fluid in the two limbs of the loop of Henle. Osmotic pressure in the renal medulla rises to a maximum of 1200 mOsm l^{-1} at the tip of the loop.

As filtrate moves up the ascending limb of the loop of Henle towards the cortex, pumping of ions without water dilutes the tubular fluid. No further changes in osmotic pressure occur in the distal tubule, but when the fluid descends the collecting duct it passes through areas of increasing osmotic pressure created by the counter-current multiplier process. If the collecting duct is water-permeable, water passes out of the collecting duct and a small volume of concentrated urine is produced. When the collecting duct is impermeable to water, a large volume of dilute urine is excreted.

The permeability of the collecting duct is controlled by vasopressin (ADH): the collecting duct is permeable to water when vasopressin is present and is impermeable when the hormone is absent. The last part of the distal convoluted tubule is also sensitive to vasopressin.

Q5 Although Irene drinks a large volume of liquids, she is also producing a large volume of very dilute urine. The water contained in fluid filtered into the renal tubules is not being adequately reabsorbed by the collecting tubules of the nephron. So, to avoid dehydration, drinking has to keep pace with the water loss from the kidney. If patients with diabetes insipidus do not drink constantly, they readily become dehydrated and very thirsty. Diabetes insipidus is caused either by a deficiency in production of vasopressin or by the inability of the kidney to respond to circulating vasopressin.

Part 2

Q6 Irene was shown to have a normal concentration of vasopressin in her plasma, so her condition is not caused by insufficient hormone production. This eliminates central diabetes insipidus. However, her kidney tubules are obviously not responding to the circulating vasopressin, so the conclusion is that she is suffering from nephrogenic diabetes insipidus.

Q7 An excess of vasopressin produces a hypo-osmolar condition with excessive water retention. This greatly dilutes the sodium content of plasma and causes an overall dilution of the extracellular fluid (ECF), which can lead to tissue swelling, for example in the brain. Mental symptoms such as confusion, irritability, seizures and coma can occur when ECF sodium falls below 120 mEq l^{-1}.

Q8 Disturbances of vasopressin secretion can be caused by tumours in the hypothalamus or pituitary gland, or trauma. Excess vasopressin is secreted following some types of brain damage and by certain tumours of the hypothalamus, prostate, pancreas or bladder. Decreased release may be caused by lesions in the posterior pituitary following inflammation or trauma.

Q9 Thiazides are considered first-line drugs in the treatment of hypertension in older people. They are also used in mild heart failure and to inhibit kidney stone formation in hypercalciuria, in addition to their use in treatment of nephrogenic diabetes insipidus.

Q10 In diabetes insipidus, reabsorption of water in the distal nephron is impaired and a large volume of dilute urine is therefore produced.

All the glucose contained in fluid filtered at the glomerulus is normally reabsorbed in the proximal tubule of the nephron. In untreated diabetes mellitus a high blood glucose concentration develops, which results in more glucose being filtered into the nephron than can be reabsorbed in the proximal tubule. Since some glucose is not reabsorbed, it remains in the filtered fluid to exert an osmotic effect, taking additional water with it through the nephron. Untreated diabetic patients are usually thirsty, drink large quantities of liquid and therefore produce a large quantity of dilute urine. However, the latter will contain glucose whereas, in diabetes insipidus, glucose is absent from the urine.

Key Points

- The body gains water via food and fluid intake plus the metabolic production of water. Routes of water loss include urine, sweat, faeces and insensible losses via the skin and lung.
- Urinary excretion of water is regulated by ADH /vasopressin, produced in the hypothalamus and released from the posterior pituitary gland. ADH acts on the distal nephron to make this area water-permeable and to allow reabsorption of water. In the absence of ADH the distal nephron is impermeable to water and dilute urine is produced.
- Concentration of urine is made possible by generation of osmotic gradients in the renal medulla. This involves a counter-current process which pumps ions without water from the loops of Henle into the medullary interstitial fluid. When the filtrate passes down the collecting ducts lying parallel to the loops of Henle, water passes out of the ducts along an osmotic gradient, provided ADH is present. A hypertonic urine is therefore produced.
- In the absence of ADH or when the nephron is unresponsive to ADH, only hypotonic urine can be produced. Large volumes of dilute urine are produced, a condition called *diabetes insipidus.* This condition can be treated with thiazide diuretic agents.

7
Blood disorders

CASE STUDY 37 An exhausted mother

Part 1

Q1 From the blood test, Maria appears to be anaemic.

Q2 Amino acids are needed to produce the plasma membrane; B_{12} (cyanocobalamine), which is stored in the liver, is required for synthesis of DNA. Folic acid is also needed for synthesis of DNA (it is a component of thymine, adenine and guanine) and for RNA synthesis. Other B vitamins are required for haem synthesis and oxidative metabolism. Iron is required in ferrous form for haemoglobin synthesis; vitamin C helps maintain iron in its ferrous form.

Q3 Macrophages of the reticuloendothelial system break down old red blood cells (RBCs).
Amino acids from cell membranes and the globin from haemoglobin are recycled.
The porphyrin constituent of haem is reduced to bilirubin, transported to the liver and excreted in the bile as a glucuronide. Iron is recycled, transferred in the blood as transferrin and stored in tissues as ferritin.

Q4 Five main types of leucocytes (white blood cells, or WBCs) can be identified. They are classified into two main subgroups, namely granulocytes, which include neutrophils, eosinophils and basophils, and agranular cells, which include lymphocytes and monocytes.

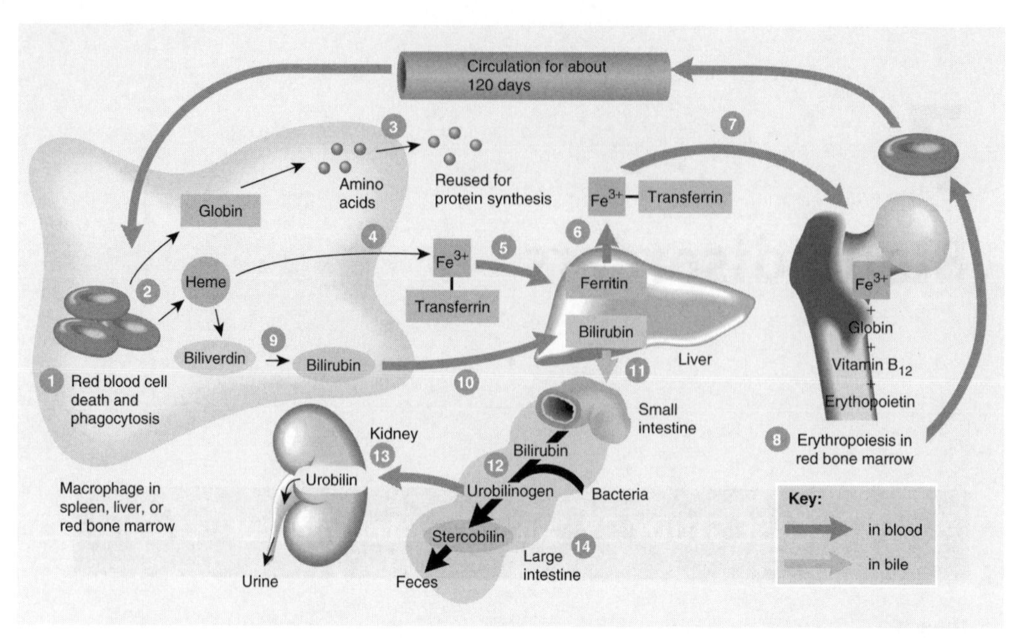

From Tortora and Derrickson *Principles of Anatomy and Physiology, Eleventh Edition 2006.* Reproduced with permission of John Wiley & Sons, Inc.

Neutrophils form 60–70% of the leucocytes in blood. They are phagocytic and their function is to remove debris and infectious particles. Eosinophils are also phagocytic and form 1–4% of white cells. They are involved in defence against parasites and in allergic responses. Basophils form 0.5% of leucocytes, are phagocytic and release histamine and other substances involved in allergic responses.

Of the agranular cells, lymphocytes form 25–30% of circulating leucocytes. They are important in producing antibodies and in immune defence. Monocytes (2–5% of leucocytes) enter tissues to become macrophages. They are phagocytic cells which are involved in defence against infectious agents and toxins, inflammation and in the reticuloendothelial system (see answers to Q3).

Q5 If anaemia develops very gradually, there may be few symptoms since compensatory cardiovascular changes occur; these ensure that oxygen supply to the tissues is maintained. Symptoms which do occur may be non-specific, for example fatigue, headache, faintness and breathlessness.

Patients with more pronounced anaemia may report tachycardia, palpitations and angina during exercise.

Anaemias may be defined in terms of a decrease in oxygen-carrying capacity of blood, a reduction in circulating RBCs or a decrease in quality or quantity of haemoglobin. They can be classified according to the appearance of the RBCs or to the cause of the condition.

Normocytic, normochromic anaemia (normal size, normal haemoglobin content) can be caused by damage to the bone marrow or by blood loss.

Macrocytic (or megaloblastic), normochromic anaemia (large cells, normal amount of haemoglobin) is due to deficiency of folic acid or B_{12}, or both.

Microcytic, hypochromic anaemia (small cells, small amount of haemoglobin) is the most common type and is due to iron deficiency.

Classification according to cause includes, for example, aplastic anaemia, which is due to bone marrow damage; haemorrhagic anaemia due to blood loss; haemolytic anaemia due to damage to red cell membranes; iron deficiency anaemia due to lack of iron; pernicious anaemia due to deficiency in B_{12} and so on.

Q6 Examination of a blood film is a routine diagnostic test. Changes in types and numbers of the cells can give important information on the patient's condition.

A differential WBC count allows the numbers of the different types of WBCs to be compared. There are usually between 5000 and 10 000 WBCs mm^{-3}.

Increased WBC levels occur in chronic and acute infection, inflammatory conditions and following tissue damage, for example burns.

Increased numbers of neutrophils can indicate bacterial infection, rheumatoid arthritis or a type of leukaemia. An increase in the numbers of lymphocytes occurs in lymphocytic leukaemia and decreased levels are found in AIDS and following steroid therapy.

Basophil numbers are increased in inflammatory and decreased in hypersensitivity reactions.

Part 2

Q7 Maria has reported recent colds and her notes show that she has a mild form of rheumatoid arthritis. Both these conditions increase the number of circulating WBCs.

Q8 This is unlikely as Maria appears to have a well-balanced food intake and is not trying to diet. In iron-deficiency anaemia, whether caused by poor dietary intake of iron or haemorrhage, RBCs are small. New RBCs entering the circulation are microcytic and carry reduced amount of haemoglobin (hypochromic). The small cells can be visualized on a standard blood film. Premenopausal women are especially likely to suffer from iron-deficiency anaemia following menstrual blood loss and childbirth. However, the blood tests show that Maria's red cells are larger than normal, so she is not suffering from this form of anaemia.

Deficiency of B_{12} or folic acid, or both, causes a macrocytic, normochromic anaemia. Maria's red cells are larger than normal, so it is probable that she has this form of anaemia. Folate deficiency is more common that B_{12} deficiency because there is usually a good store of B_{12} in the liver. When B_{12}

or folate levels are low, cell division and the maturation of RBCs is reduced. Haemoglobin synthesis is not adversely affected but, since B_{12} is required for DNA synthesis, rapidly dividing cells such as those in bone marrow are susceptible to B_{12} deficiency. A smaller number of cells are produced and the overall haemoglobin concentration of blood falls.

Patients with B_{12} deficiency often have problems absorbing the vitamin. The gastric mucosa secretes an intrinsic factor which binds B_{12} to form a complex, allowing absorption in the ileum. When secretion of intrinsic factor is diminished, B_{12} absorption is reduced, for example in gastric cancers, inflammatory states, trauma or surgery.

Lack of B_{12} can cause neurological problems as it is also involved in myelination of axons. Patients may show mood swings and appear to suffer more minor infections and gastrointestinal upsets than normal.

Folic acid is involved in DNA synthesis and is needed to form three of the four bases of DNA. It is absorbed in the upper small intestine, but this does not require intrinsic factor. Folate deficiency may occur in alcoholics and other chronically malnourished people.

WBCs are also affected in B_{12} or folate deficiency; the neutrophils are abnormal and show increased segmentation of their nuclei.

Q9 B_{12} is ineffective when given by mouth if there is deficiency of intrinsic factor, as it would not be absorbed. It must be given as a depot injection, which lasts a few months. Folate supplements can be given by mouth. However, if the patient has neurological symptoms, folic acid supplements alone are not adequate: B_{12} must also be administered. So the two forms of therapy are not equally effective.

Q10 Maria appears to be suffering from pernicious anaemia, which is due to failure of B_{12} absorption. In her case the underlying problem may be her rheumatoid arthritis, a chronic autoimmune, inflammatory condition. In autoimmune diseases the immune system attacks and damages normal tissues, including both joints and the stomach mucosa, which produces the intrinsic factor needed for B_{12} absorption.

Important factors in this case are: the tiredness, low RBC count, haematocrit and haemoglobin, which are consistent with anaemia. Her previous diagnosis of rheumatoid arthritis, her mood changes and weakness together with the high erythrocyte sedimentation rate (ESR) is consistent with her autoimmune disease. Autoimmune conditions are accompanied by inflammation and could damage her gastric mucosa. The type of anaemia is deduced from the high mean red cell volume together with results from the blood test, which confirm that the type of anaemia is macrocytic (megaloblastic) and normochromic. Such anaemias are known to be associated with B_{12} and folic acid deficiency.

Maria was taking a non-steroidal anti-inflammatory drug (NSAID), which has both an analgesic and an anti-inflammatory action, but causes gastric irritation, which can lead to asymptomatic blood loss in some patients. It is possible that Maria has suffered some gastric irritation and mild blood loss, but the results of the blood test show that her anaemia was not associated with haemorrhage, but was of another type. Her megaloblastic anaemia is associated with B_{12} and folate deficiency.

Q11 The ESR is a simple, non-specific indicator of inflammation. In inflammatory conditions, like rheumatoid arthritis, there may be high values, but the ESR cannot be used to identify the type or extent of inflammation.

So the ESR is not diagnostic in itself, but is an indicator that there is an ongoing inflammatory process in the body. It is not useful in diagnosing the presence or type of anaemia but, taken together with the previous diagnosis of rheumatoid arthritis and the blood test results, it helps to support the diagnosis of pernicious anaemia.

Key Points

- Erythrocytes are produced in the red bone marrow and survive in the circulation for approximately 120 days. Old or abnormal red cells are destroyed by the reticulendothelial system and the products recycled. The amino acids liberated enter the liver; haem is broken down to ferrous iron and the pigment bilirubin. Iron is stored as ferritin in tissues, and bilirubin is excreted in bile as a glucuronide.
- There are five types of leucocytes. Neutophils, basophils and eosinophils possess granular cytoplasm. These cells are involved in defence against bacteria, viruses and parasites. The agranular cells are the lymphocytes and the phagocytic monocytes. Lymphocytes produce immunoglobulins; monocytes enter tissues to become macrophages; they are involved in inflammation and in defence against infectious agents.
- Examination of a stained blood film allows a differential WBC count to be performed. This is useful for diagnosis as characteristic changes in the proportion of the different WBCs are observed in different diseases.
- Anaemias are classified according to the size and haemoglobin content of erythrocytes or to the cause of the condition. In the latter classification bone marrow damage causes aplastic anaemia; haemorrhagic and haemolytic anaemia are due to blood loss or damaged red cell membranes respectively, iron deficiency and pernicious anaemia are due to deficiency of iron and vitamin B_{12} respectively.
- The ESR is a useful, non-specific indicator of inflammation. It cannot indicate the type or severity of inflammation.

CASE STUDY 38 Patsy's Australian journey

Part 1

Q1 Blood clots (thrombi) which form in the venous part of the circulation are associated with slow or sluggish venous blood flow. This condition, called *deep-vein thrombosis* (DVT), has long been recognized as a risk factor for people immobilized by extended periods of bed rest and for passengers on long journeys with little room to move their legs. Sometimes the thrombi are detached from the vessel wall and travel to other parts of the circulation, causing serious obstruction to blood flow. A thrombus which detaches and enters the pulmonary circulation is particularly serious.

Q2 Patsy is at risk of DVT because of slow venous blood flow in her legs during a long flight in cramped conditions with little room to move her legs and feet, pressure of the seat edge on her legs and no opportunity to move around the aircraft. Her risk is increased by other factors, such as:

- recent pregnancy and childbirth
- dehydration because of the hot dry atmosphere in aircraft and inadequate fluid intake
- smoking.

In other passengers DVT risk may be increased by:

- family history of DVT
- recent surgery
- obesity
- oral contraceptives and certain cancers which can alter levels of clotting factors in the blood.

Q3 The thrombi which detach and enter the pulmonary circulation usually form in the deep veins of the leg, particularly in the calf muscle, or in the pelvis. Blood returns to the heart from the leg and abdominal veins against gravity. During normal body movement, the contraction of large skeletal muscles in the legs 'massages' blood upwards to the vena cava, but in prolonged inactivity this action is missing, blood flow slows and tends to pool in lower parts of the body. The seated position also involves some pressure on the legs from the edge of the seat, which further slows blood flow in leg veins. DVT has been called *economy-class syndrome* because the crowded economy passengers have little room for stretching and moving their legs. However, since DVT is associated with immobility, it occurs in any form of long-distance travel when passengers remain immobile for long periods.

In most cases the thrombi which form do not detach from the wall of the vein. But the passenger may later be aware of some local symptoms of DVT, such as swelling, pain, tenderness or redness in the leg.

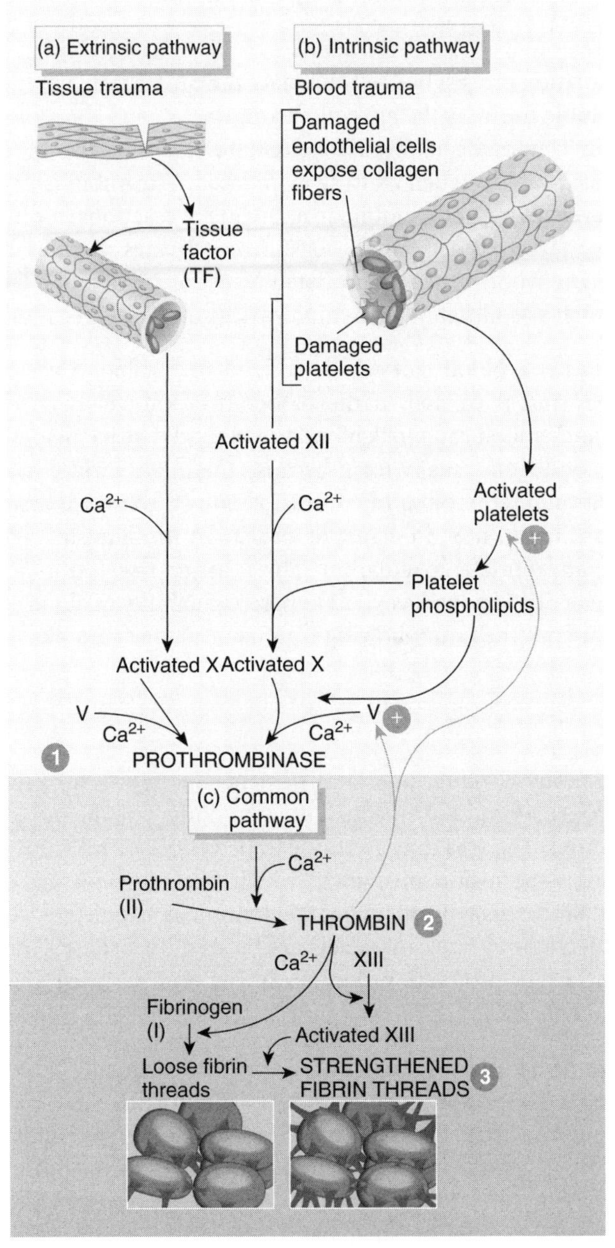

From Tortora and Derrickson *Principles of Anatomy and Physiology, Eleventh Edition 2006.* Reproduced with permission of John Wiley & Sons, Inc.

Q4 Vitamin K is a fat-soluble vitamin which is composed of a number of related compounds known as *quinones*. Most of the protein clotting factors in blood are produced in the liver and depend on the presence of adequate vitamin K for their synthesis. Some of the necessary vitamin K is present in the diet (in green, leafy vegetables), but most is synthesized by bacteria in the intestine. Vitamin K is particularly important in maintaining blood levels of coagulation factors II, VII, IX and X.

Q5 The coagulation process basically involves the conversion of soluble fibrinogen into insoluble fibrin by the enzyme thrombin.
Thrombin cannot be allowed to circulate freely in the blood, but must be produced rapidly when clotting is initiated. Thrombin is derived from its inactive precursor, prothrombin, in a cascade of reactions in which a sequence of inactive factors is activated: each factor activates the next by proteolytic cleavage. The whole process is initiated by exposure of blood to an abnormal surface, such as collagen, which initially triggers the aggregation of platelets.

Q6 A *thrombus* is a blood clot which is fixed to the blood vessel wall. When it detaches and is carried in the blood, it is known as an *embolus*. Both thrombi and emboli can block blood vessels and deprive tissues of oxygen. In arteries blood clots usually form because the inner surface has been altered by deposition of atheroma. In contrast venous thrombosis results from slow or stagnant blood flow in veins, or defects in mechanisms which normally oppose inappropriate coagulation. Three major risk factors for pulmonary embolism are (i) venous stasis, (ii) hypercoagulability of blood and (iii) injury to vascular endothelium following trauma or plaque rupture.

Part 2

Q7 If the embolus is quite large and obstructs a significant area of the pulmonary circulation, the affected area of lung will be underperfused or non-perfused. The area may continue to be ventilated for some time, causing a ventilation–perfusion mismatch, which leads to poor gas exchange and abnormal blood gas tensions. The lung volume in the affected area decreases, and this decrease in size can sometimes be seen on a chest X-ray. After some hours, surfactant production declines in the non-perfused area of lung and the alveoli collapse.

Q8 A number of natural substances act as anticoagulants in blood and prevent intravascular clotting in normal vessels. The normal vascular endothelium discourages the first stage of thrombus formation by inhibiting platelet aggregation. It produces nitric oxide and prostacyclin, both of which are anti-aggregatory. In addition, antithrombin III is present in blood. It is a circulating protease inhibitor which blocks the activity of several clotting factors. Circulating heparin also inhibits many components of the coagulation cascade.

The breakdown, or lysis, of formed clots depends on production of a proteolytic enzyme, plasmin, from the inactive precursor plasminogen, when coagulation is initiated.
Plasmin lyses fibrin in the formed thrombus.

Q9 Fibrinolytic drugs, such as streptokinase, are given intravenously to lyse clots in the pulmonary circulation and coronary circulation. Occasionally, the thrombotic mass must be removed surgically. Streptokinase activates plasminogen to form plasmin, which degrades the fibrin in the thrombus. Heparin may be given intravenously to prevent further coagulation.

Q10 In high-risk patients scheduled to have general surgery, subcutaneous heparin can be given to reduce the risk of DVT and pulmonary embolism occurring after the operation. If DVT or pulmonary embolism has already occurred in hospital patients, the immediate treatment normally includes intravenous heparin.
When these patients are discharged from hospital, prophylactic treatment with an oral anticoagulant is recommended to prevent recurrence of the thrombosis. Warfarin sodium, which antagonizes the effects of vitamin K, is used in prophylaxis and treatment of DVT and pulmonary embolism. It is usual to start with an induction dose of 10 mg daily for two days: the dose can then be reduced. Patients need to be monitored as there is a risk of haemorrhage with oral anticoagulant drugs.

Q11 Although low-dose aspirin can be safely used for prevention of intravascular coagulation, it is not currently licensed in the United Kingdom for prevention of travel-related DVT.
Long-haul travellers are advised to wear compression stockings, to encourage the return of blood from the legs to the heart. However, these are not suitable for patients with arterial disease.
On the flight or journey the traveller should avoid crossing their legs, regularly flex feet and toes, bend and straighten their legs at intervals during the flight and, when possible, leave their seat to move around. They are advised to keep well hydrated by drinking water or fruit juices but to avoid alcohol and caffeine, as they tend to increase dehydration, which is a risk factor for DVT.

Key Points

- Intravascular coagulation can occur in the deep veins of the leg during long journeys, particularly in cramped conditions. Risk factors for DVT include: recent pregnancy and childbirth, dehydration, recent surgery, obesity, oral contraceptives, certain abdominal cancers and family history.

If the thrombus dislodges and blocks a large pulmonary artery, chest pain and collapse may occur.

- Most factors associated with clotting are produced in the liver and require the presence of vitamin K. The coagulation process involves producing thrombin, which converts soluble fibrinogen to insoluble fibrin.
- Anticoagulants also exist in blood; these include heparin and antithrombin III, which is a circulating protease. If small clots form in the circulation, they are dissolved by plasmin, formed from the inactive precursor plasminogen.
- Hospital patients at risk of inappropriate clotting may be given intravenous heparin and prescribed oral warfarin on discharge. Fibrinolytic agents such as streptokinase are used therapeutically to lyse clots formed in the coronary or pulmonary circulation.
- Low-dose aspirin is used in prevention of intravascular coagulation but is not licensed for prevention of travel-related DVT.

CASE STUDY 39 The dizzy blonde

Q1 *Anaemia* is defined as a reduction in RBC mass with a haematocrit of less than 37% in females. Lizzie appears to be suffering from iron-deficiency anaemia.

Q2 Lizzie has experienced four pregnancies in quick succession. Premenopausal women are especially likely to suffer from iron-deficiency anaemia, because of menstrual blood loss (2–3 mg per day) and childbirth (500–1000 mg per pregnancy). She has recently greatly decreased her food intake and in addition her sore tongue has made it difficult to chew meat, which would limit her intake of haem iron from red meat, a rich source of available iron in the diet. All these factors, loss of iron via pregnancy and childbirth together with reduced dietary intake, contribute to iron-deficiency anaemia. In addition Lizzie has other clinical features consistent with iron-deficiency anaemia: she is dizzy and breathless, has experienced palpitations, shows angular stomatitis (red areas at the corner of the mouth), glossitis (inflammation of the tongue) and dry, brittle hair. Brittle nails may also be observed in iron deficiency.

Q3 Both the low haemoglobin content of the blood and the reduced haematocrit are consistent with anaemia. The low mean corpuscular volume shows the red cells are small, a characteristic feature of iron-deficiency anaemia. The low concentration of ferritin, which reflects the amount of iron stored in the body, is also consistent with a deficiency of iron in the body.

Q4 When the supply of iron is greatly diminished, haemoglobin synthesis is restricted. Erythropoiesis continues and is controlled by erythropoietin from the kidney. Release of this hormone is increased in anaemia, in response to a reduced concentration of circulating haemoglobin. The bone marrow will then be stimulated to produce more red cells, but of smaller size and with smaller haemoglobin content: a blood film may show red cells of unequal sizes; this is known as *anisocytosis*.

Q5 The average diet in the United Kingdom contains approximately 15 to 20 mg of iron; only about 10% of this is absorbed. The main iron content of the diet is haem iron derived from the haemoglobin and myoglobin in red meat, and this is the form of iron which is most readily absorbed. Non-haem iron, for example that derived from cereal products, is less well absorbed. The majority of iron absorption occurs in the duodenum and first part of the jejunum. Non-haem iron absorption is very variable, and some types of food, such as bran and egg yolk, limit its absorption. Gastric acidity helps to keep iron in the ferrous state, which is more easily absorbed than the ferric form. Iron absorption is increased when iron stores are low and when there is increased erythropoietic activity.

Q6 At birth RBCs are produced in all bones, but in adults it is confined to the marrow at the end of the long bones plus that of the pelvis, sternum, ribs and cranium. RBCs (and all peripheral blood cells) are derived from a pluripotential stem cell in the marrow, which differentiates, makes haemoglobin and eventually loses its nucleus and other organelles to become the mature form. Control of RBC production is via erythropoietin, a glycoprotein, which is produced in the kidney and stimulates the pluripotential stem cell to begin differentiating to form RBCs. Following a series of divisions, the cells lose their nucleus and become the familiar biconcave red cells. Young, newly formed RBCs retain some ribosomes and mitochondria for a time and are known as *reticulocytes*. Reticulocytes normally form 1% of the RBCs, but when a person is recovering from anaemia or haemorrhage or suffers increased red cell destruction the number of reticulocytes in the circulation increases, and this is a useful measure of erythropoiesis.
Ageing erythrocytes are destroyed by the mobile phagocytic macrophages of the reticuloendothelial system, mainly in the spleen. The average lifespan of a red cell is about 120 days.

Q7 Iron-deficiency anaemia, can be caused by loss of blood following haemorrhage, poor dietary intake of iron, bone marrow damage, increased destruction of red cells or increased demand for iron (which can occur during growth or pregnancy) and poor/reduced absorption of iron from the intestine. Iron deficiency results in the production of small RBCs, and haemoglobin synthesis is insufficient. The RBCs entering the circulation are microcytic and carry a smaller than normal amount of haemoglobin (hypochromic). The small red cells can be visualized on a standard blood film. Premenopausal women are especially likely to suffer from iron-deficiency anaemia because of menstrual blood loss and childbirth.

Q8 Since iron deficiency may follow pathological blood loss, some investigation of possible sources of blood loss, particularly into the gut, is advised before treatment of the anaemia is started. Haemorrhage into the gut, for example from erosion or ulceration of the stomach or from sites in the large intestine in colorectal cancer, can be simply detected by testing for (occult) blood in the faeces.

Q9 Treatment consists of correcting the underlying cause and replacing the deficient iron. Oral administration of ferrous sulphate (containing 120–200 mg of elemental iron daily) is the standard treatment. This should be taken on an empty stomach, if tolerated, with vitamin C (ascorbic acid), which aids absorption. When the normal haemoglobin concentration in blood has been achieved, the treatment should be continued for an extra three to six months to ensure that body iron stores are replenished.

Q10 The most common side effect of iron treatment is gastrointestinal irritation. There may be nausea, diarrhoea or constipation and epigastric pain;

constipation can be a particular problem in elderly patients. The side effects may be minimized if the dose is reduced or another iron salt, such as ferrous gluconate, substituted. Modified release preparations (polysaccharide–iron complexes) may reduce gastrointestinal irritation, but are more expensive and provide little therapeutic advantage.

Q11 Acute iron toxicity causes nausea, vomiting, abdominal pain, diarrhoea and bleeding from the gut, because of the corrosive effects of iron salts on the gastrointestinal epithelium. If there is evidence of poisoning, the stomach is washed out (gastric lavage) as soon as possible and the patient is given a chelating agent to complex the iron and promote its excretion: the agent used is desferrioxamine.

Key Points

- Iron-deficiency anaemia is common in premenopausal females and is associated with iron loss in pregnancy, childbirth and menstruation and in nutritional iron deficiency. Anaemia reduces the oxygen-carrying capacity of blood.
- Symptoms of anaemia include dizziness, weakness and breathlessness. There may also be skin and hair changes, glossitis and stomatitis. Patients may become breathless and experience irregular heart beats.
- Iron absorption occurs in the small intestine, mainly the duodenum and early jejunum. Haem iron from meat is more easily absorbed than non-haem iron from other food products. Ferrous iron is easier to absorb than ferric iron and absorption is increased when iron stores are low.
- Production of RBCs is controlled by erythropoietin from the kidney. Erythropoiesis occurs mainly in red bone marrow of the long bones, sternum and pelvis of the adult. Newly formed red cells enter the circulation as reticulocytes, recognizable because of the inclusion of ribosomes and mitochondria in their cytoplasm.
- A common therapeutic iron salt is ferrous sulphate, which may cause gastrointestinal irritation, constipation or diarrhoea in some patients. In acute poisoning with iron there may be additional symptoms of pain and bleeding from the gut. Acute iron toxicity is treated by washing out the stomach and administering an iron-chelating agent, desferrioxamine.

8

Gastrointestinal disorders

CASE STUDY 40 Mr Benjamin's bowel problem

Q1 Faecal material usually remains in the colon for about 24 hours, but the rectum is normally empty. There are generally slow mixing and propulsive contractions in the colon but mass movements of colonic content also occur, usually after a meal, when strong contractile waves push the content into the rectum and distend it. Distension stimulates sensory receptors in the rectum and initiates the defecation reflex, a reflex involving parasympathetic nerves in the sacral spinal cord, together with conscious awareness of the urge to defecate. At the same time the smooth muscle of the internal anal sphincter is relaxed and the somatic nerves supplying striated muscle in the external anal sphincter are inhibited, allowing the sphincter to relax. Voluntary control of defecation is learnt in early childhood and involves voluntary contraction of the external anal sphincter.

Q2 Constipation is a condition in which faecal material moves too slowly through the large intestine. As a result too much water is reabsorbed; hard, dry faeces which are difficult to move and very abrasive are produced. Infrequent or difficult defecation is a common problem in the elderly as ageing is associated with a decline in both secretory activity and motility in the gut. Constipation could develop because of emotional problems, inactive or sedentary lifestyle, lack of fibre and fluid in the diet, intestinal muscle weakness, a neurogenic disorder or an iatrogenic effect. Iatrogenic conditions are those caused by drugs or other medical treatments.

Q3 Mr Benjamin eats little fruit or vegetables and is therefore likely to have inadequate fibre (roughage) in his diet; he also drinks very little fluid. This

Clinical Physiology and Pharmacology Farideh Javid and Janice McCurrie

results in a small volume of faecal material moving through the colon. Slow passage of the material through the colon favours fluid reabsorption: excessive drying and compaction of the faecal material may then occur. Mr Benjamin is also inactive, taking only occasional short walks, which also increases the likelihood of constipation.

Q4 Constipation can be a troublesome side effect of opiates used for pain relief, for example morphine and codeine. It is also a side effect of some calcium channel blocking agents, antacids containing aluminium compounds and iron salts used in the treatment of anaemia.

Q5 Laxatives are used to treat constipation. They change the consistency of the faeces, increasing the frequency of defecation by accelerating the rate of faecal passage through the colon and elimination of stool from the rectum. There are four main types of laxative: bulk-forming preparations, such as sterculia and ispaghula; hyperosmolar or saline solutions, such as magnesium sulfate; faecal softeners/wetting agents, such as docusate (dioctyl sodium sulfosuccinate); and stimulant or irritant laxatives, such as senna and bisacodyl. Before a laxative is prescribed, it is important to ensure that the patient really is constipated, as the frequency of normal defecation varies considerably between patients, ranging from three times a day to one defecation every three days. Constipation may be secondary to another, possibly serious, condition such as intestinal obstruction, and this should be excluded before treatment begins. In general, a bulk-forming or hyperosmolar laxative is tried before stimulant compounds are used.

Q6 Lactulose is a hyperosmotic liquid containing a disaccharide of galactose and fructose which is not absorbed from the intestine. The recommended dosage is 15 ml twice a day. It passes unchanged into the colon and produces an osmotic effect, directing fluids into the colon content, which expands the bowel and initiates peristalsis. Production of lactic and acetic acids from lactulose is brought about by bacteria in the large intestine, which in turn further stimulates peristalsis. Lactulose is safe for diabetic patients since the sugar content is not absorbed.

Q7 Laxatives are often misused/abused, for example in slimming disorders, to increase gut transit rate and so limit absorption of foods. Side effects which may occur include: flatulence, and abdominal distension or discomfort with bulk-forming and osmotic laxatives. Other adverse effects may include: diarrhoea, nausea, vomiting, weakness, dehydration and electrolyte imbalances, for example hypokalaemia. The most prominent side effect of the powerful stimulant/irritant laxatives is abdominal cramping, which is due to increased peristalsis.

Q8 Mr Benjamin has a restricted, low-roughage diet and his colonic motility will be helped by increasing its fruit and vegetable content. It may be possible to find him a day care place at which a healthy, balanced meal can be provided; he will also benefit from the exercise and social interaction involved. It is important that Mr Benjamin maintains his fluid intake during the day, to avoid dehydration and promote colonic transit.

Key Points

- Constipation is associated with slow transit of faecal material through the large intestine and increased fluid absorption, resulting in hard, dry faeces.
- Constipation is common in the elderly, in people with emotional problems or those with an inactive/sedentary lifestyle, and also with lack of fibre and fluid in the diet, intestinal muscle weakness and neurogenic disorders.
- Constipation can be associated with the use of opiates, such as morphine and codeine, calcium channel blocking agents, antacids containing aluminium compounds and iron salts used in the treatment of anaemia.
- Drug treatment involves the use of laxatives. There are four main types: bulk-forming preparations, hyperosmolar or saline solutions, faecal softeners/wetting agents and stimulant or irritant laxatives.
- Adverse effects of laxative use or misuse include: flatulence, abdominal distension, cramps and discomfort, diarrhoea, weakness, dehydration and electrolyte imbalances.

CASE STUDY 41 A disturbed holiday

Q1 A basic definition of *diarrhoea* is an increase in frequency and/or volume of the faeces: in an adult the average quantity of faeces passed per day is 200 g. Ninety-nine percent of the fluid ingested each day plus the gastric and intestinal fluid secreted into the gut (7–8 l daily) is usually reabsorbed in the intestine, and only 150 ml is excreted via the faeces.

There are three underlying causes of diarrhoea: osmotic, secretory and motility.

Osmotic diarrhoea occurs when a non-absorbable substance draws fluid into the intestine by osmosis, for example lactase deficiency, when unabsorbed lactose remains in the intestine. This type of problem also occurs in malabsorption disorders, for example in celiac disease.

Secretory diarrhoea may be caused by excessive secretion of fluid and electrolytes into the intestinal lumen as a result of a bacterial toxin or a tumour producing a secretory stimulant.

An increase in intestinal motility also causes diarrhoea; because of very rapid transit of material through the gut, there is insufficient time to fully reabsorb gastric and intestinal secretions. Causes of motility change include impaired autonomic control, for example in peripheral neuropathy of diabetes or following some surgical procedures which shorten the intestine.

Q2 Diarrhoea can be acute or chronic. Acute diarrhoea has a sudden onset and, if it is due to a viral agent, usually lasts 24–48 hours. Acute diarrhoea may also be due to unwise food consumption or to food poisoning. Traveller's diarrhoea, which affects people travelling outside their own countries, usually lasts two to five days. A working definition of the latter type of diarrhoea is three or more unformed stools in 24 hours and at least one other symptom, such as: faecal urgency, fever, nausea, vomiting, abdominal pain or cramps. Chronic diarrhoea may be due to: enteric infection with parasitic or fungal organisms, drugs, malabsorption or inflammatory bowel disease. If the diarrhoea is severe, the major problem is loss of fluid and electrolytes which results in severe dehydration and electrolyte imbalance. This can be a particular problem in infants, young children and the elderly.

Q3 The most common agent is *Escherichia coli*, but other bacterial causes include: *Campylobacter jejuni* and *Salmonella* species. A minority of cases appear to involve viral infection, such as rotavirus.

Q4 Mrs Kaye may be at risk because this was her first visit abroad and possibly she was particularly affected by a change in diet or did not take the usual precaution of drinking bottled water and avoiding salads and unwashed fruit. Mrs Kaye also takes ranitidine for indigestion and there is some evidence that patients with reduced gastric acid, including those who take H_2 antagonists such as ranitidine, are at increased risk of traveller's diarrhoea.

Q5 Different treatments can be prescribed depending on the type or cause of diarrhoea. These include oral rehydration, absorbents, antimotility agents such as opioids or intestinal flora modifiers. In all cases maintaining fluid intake helps to improve symptoms.

Oral rehydration therapy consists of a mixture of salt and glucose, or another carbohydrate, in clean, preferably boiled, water. Commercial sachets of the materials are available and form the most suitable treatment for children and the elderly.

Absorbents such as kaolin are not recommended for traveller's acute diarrhoea.

The most useful antimotility agent for adults is loperamide because it has specific effects on the gastrointestinal (GI) tract. It is not recommended for children as it decreases the clearance of the pathological organism from the gut and so prolongs the problem.

Very occasionally, an antibiotic may be necessary, depending on the organism involved.

Q6 Loperamide hydrochloride is an opioid. The starting dose will be 4 mg, which can be reduced to 2 mg, three times a day for five days if necessary. Opioids act on μ opiate receptors in the myenteric plexus of the intestine and may modulate acetylcholine release to reduce peristalsis. They trigger mucosal transport of ions and water out of the lumen and cause a reduction in secretion. The absorption of fluid and electrolytes is increased since the stool remains in the colon for a longer period. Loperamide does not produce sedation or other central effects associated with opiates, since it does not cross the blood–brain barrier.

Q7 Intestinal flora modifiers include *Lactobacillus acidophilus* or *Lactobacillus bulgaricus*. They help to establish and maintain the balance of the intestinal flora by enhancing or replacing the normal flora. Millions of bacteria normally exist in the gastrointestinal tract, particularly in the colon. In the colon, these bacteria help to produce vitamins K, B_{12}, thiamine and riboflavin, and also digest small amounts of cellulose, producing gases. Following administration of antibiotics or after diarrhoea, the normal flora of the intestine may be reduced or changed, which can then lead to secondary diarrhoea and gas production. The presence of 'good' bacteria such as *Lactobacillus* spp helps to prevent the growth of unfavourable bacteria in the colon.

Key Points

- Acute diarrhoea has a sudden onset and, if it is due to a viral agent, usually lasts 24–48 hours. It may be due to unwise food consumption, food poisoning or an infectious agent such as a virus.
- Traveller's diarrhoea, which affects people travelling outside their own countries, usually lasts two to five days. It involves three or four unformed stools in 24 hours and at least one other symptom, such as: faecal urgency, fever, nausea, vomiting, abdominal pain or cramps.
- Chronic diarrhoea may be due to: enteric infection with parasitic or fungal organisms, drugs, malabsorption or inflammatory bowel disease. In severe cases it can lead to severe dehydration and electrolyte imbalance. This can be a particular problem in infants, young children and the elderly.
- The most frequent cause of traveller's diarrhoea is *E. coli* but other bacterial causes include: *C. jejuni* and *Salmonella* species. A minority of cases appear to involve viral infection, such as rotavirus.
- Therapies include oral rehydration, absorbents, antimotility agents such as opioids or intestinal flora modifiers. In all cases maintaining fluid intake helps to improve symptoms. Very occasionally, an antibiotic may be necessary, depending on the organism involved.

CASE STUDY 42 Jude's sudden admission to the hospital

Part 1

Q1 Because of her symptoms and the illness suffered by Jude's boyfriend, a tentative diagnosis of infectious hepatitis is initially made.

Part 2

Q2 No. Jude showed no signs of jaundice, such as a yellow tinge to the skin. In addition, her scans, which could have indicated a problem with her liver or gall stones, revealed a normal liver and bile duct.

Q3 Signs of jaundice: jaundice gives a yellowish colour to the skin and mucous membranes, usually easiest to see in the cornea. The yellow colour is due to the presence of breakdown products of haemoglobin such as bilirubin in tissues, which the liver usually removes from the blood. Jaundice is indicative of liver disease, obstruction of the bile ducts or haemolytic disease. Bilirubin stains not only the tissues but also all body fluids, including plasma and urine, and the patient's urine can become really dark.

Part 3

Q4 The revised diagnosis is pancreatitis as the patient did not appear to be jaundiced and a specimen of urine showed a normal colour. Her pain and other symptoms are consistent with pancreatic inflammation.

Q5 Nothing is given by mouth, to minimize stimulation of the pancreas. Secretion of fluid and enzymes from the pancreas is stimulated by the presence of chyme in the small intestine. Fluids given by mouth will enter the duodenum and stimulate pancreatic secretion; this must be reduced to a minimum in order to reduce further irritation of the inflamed pancreas and to 'rest' the tissue. The fluids which will be needed to replace water lost through vomiting are therefore given by the intravenous route. In some patients gastric fluids secreted by the chief and parietal cells of the gastric mucosa are aspirated from the stomach, to ensure nothing passes into the duodenum.
Glucose and saline are administered intravenously to maintain blood glucose in the normal range, to ensure that the patient is adequately hydrated and has a urine output.

Q6 Alpha-amylase (α-amylase) is concerned with the digestion of starch to disaccharides and other products in the gut. The salivary glands also produce α-amylase, the parotid glands producing the greatest amount.

Q7 Pancreatic exocrine tissue produces amylase, lipase and a range of serine proteases, enzymes such as trypsin (which is also elevated in pancreatitis), chymotrypsin and elastase, also nucleases, carboxypeptidase and aminopeptidase.

Q8 The normal pancreas is not damaged by the enzymes it produces because they are produced and stored in an inactive form, for example trypsinogen and chymotrypsinogen.
When trypsinogen enters the small intestine, it is converted to trypsin by enterokinase. The trypsin produced then converts chymotrypsinogen and other proteolytic enzymes to their active form.

Q9 Actions of α-amylase on gut content: α-amylase hydrolyses the 1,4-α-glycosidic bonds of glucose polymers in starches. The products are the disaccharide maltose and oligosaccharides. Digestion to monosaccharides is completed when the chyme contacts the intestinal mucosa, as maltase is present in the brush border of these cells. Maltase converts maltose to glucose. A large amount of starch is consumed in the average diet, so amylase is an important digestive enzyme.
The low concentration of amylase normally found in the blood has been produced in small amounts by many tissues, but the greatest amounts are produced by the pancreas and the parotid salivary gland.

Q10 Normally, pancreatic secretion is stimulated by eating. The factors involved are both nervous and hormonal. Pancreatic secretion is increased by vagal (parasympathetic) stimulation and inhibited by sympathetic stimulation. Cholecystokinin (CCK), released from the wall of the duodenum, stimulates pancreatic juice, which is rich in enzymes. Secretin (the first hormone to be discovered, by Bayliss and Starling), which is also produced in the duodenum, stimulates a pancreatic fluid with a high bicarbonate content.

Q11 The most useful tests in suspected pancreatitis are biochemical: the tests involve measurements of the concentration of amylase and lipase in blood. Increased blood levels of α-amylase can be found in a number of other conditions, for example disease of the ovaries, but elevated amylase is most often seen in acute and chronic pancreatitis. During acute attacks, the blood can contain three to five times more amylase and lipase than normal and the pattern of change is characteristic.

Q12 Endocrine secretions are hormones, substances made in a location from which they are carried in blood to their site of action elsewhere. Exocrine secretions are produced and secreted into another organ or onto the body surface, usually via ducts which carry them to their site of action.

Q13 The pancreas produces hormones in areas of endocrine tissue: the islets of Langerhans. The major hormones are insulin, from the beta-cells (β-cells), and glucagon, from alpha-cells (α-cells). Delta cells produce somatostatin.

Q14 Excessive alcohol consumption damages the pancreas. Patients usually develop pancreatitis after many years of excessive alcohol consumption, but occasionally it can occur after only one year of heavy drinking. There is no safe level of alcohol intake, below which no damage to the pancreas can occur: even moderate or social drinkers are at risk. Some people have more than one attack of pancreatitis and recover well from each, but in a few individuals acute pancreatitis can be severe, and a life-threatening illness develops.

Q15 Reasons for observing high concentrations of amylase and lipase in the blood: consumption of large amounts of alcohol appears to cause inflammation of and damage to the pancreatic exocrine cells. Some researchers think that pancreatic cells may become sensitized to alcohol and be further affected even if only very small amounts of alcohol are taken. Following secretory cell damage, the enzymes leak out into surrounding areas, irritating the secretory cells, causing further enzyme release, then oedema, haemorrhage and cell death. Trypsin and chymotrypsin appear to initiate the process and in turn activate other pro-enzymes, which damage tissues further. The large quantity of released enzymes passes into tissue fluids and into the plasma, causing the blood concentration of amylase to rise rapidly following pancreatic damage. This pattern, of rapid rise in amylase and the slower, more sustained rise in lipase, is characteristic of pancreatitis and is a simple test which aids differential diagnosis of the condition.

Q16 In pancreatitis the cells which produce enzymes are damaged and the quantity of enzymes entering the duodenum is reduced. At some point there will be an inadequate amount of enzyme to deal with carbohydrate and, particularly, with lipids in the diet. A considerable portion of the daily calorie intake will not be digested or absorbed, and fat-soluble vitamin absorption will be compromised. Patients will suffer from malabsorption, particularly of fats. In a normal individual $<7\%$ of dietary fat passes through the gut undigested. Patients with pancreatic insufficiency may excrete $>20\%$ of their dietary fat unchanged, a condition known as *steatorrhoea*. Faeces which contain large quantities of undigested fat float in water and also smell very unpleasant.

When high alcohol consumption continues year on year, a chronic pancreatitis may develop causing increasing and irreversible loss of pancreatic tissue. The secretion of enzymes into the duodenum is then permanently reduced to a very low level. Since these patients often replace meals with alcohol, their food intake is unlikely to be well balanced. The combination of inadequate digestion and absorption of nutrients with an inadequate dietary intake leads to deficiency of both specific nutrients and total calories, resulting in malnutrition and weight loss.

Patients may be prescribed an enzyme supplement, which is added to food to allow digestion of dietary fat. This both reduces the steatorrhoea and helps the patient to regain lost weight.

Key Points

- The pancreas secretes a variety of enzymes into the gut. These include: proteolytic enzymes, such as trypsin and chymotrypsin, lipase and amylase. The salivary glands also produce an amylase. The major hormones produced by the pancreas are insulin, glucagon and somatostatin.
- Inflammation of the pancreas, pancreatitis, can be a consequence of excessive alcohol intake and causes severe pain. In chronic pancreatitis the exocrine cells which produce enzymes are damaged and smaller quantities of enzyme are released into the gut so that a major portion of the diet remains undigested and is not absorbed. Patients suffer malnutrition and weight loss.
- Pancreatic secretion is controlled by nervous and hormonal factors. Both vagal stimulation and the hormones secretin and CCK, released from the duodenum, stimulate secretion. In acute pancreatitis nothing is given by mouth, in order to minimize pancreatic stimulation and to rest the pancreas. Fluid and glucose are given intravenously to maintain hydration.
- Certain biochemical changes are useful in diagnosing pancreatitis: pancreatic damage results in amylase and lipase being released into the blood. Amylase concentration rises over 3–12 hours and returns to normal in three to four days. The rise in lipase is slower and restoration of normal blood levels takes longer.
- Hepatitis could result in symptoms similar to those experienced in this case study, but would be likely to cause jaundice, a yellow skin colouration which is due to the presence of haemoglobin breakdown products in the tissues.

CASE STUDY 43 The producer's stomach ache

Q1 The stomach acts as a temporary reservoir for food and as a mixing chamber, allowing small amounts of gastric contents (chyme) to enter the duodenum at intervals. The acid environment and mechanical activity in the stomach starts the breakdown of food items and the acidity of the stomach eliminates many infectious organisms present in ingested material. Finally, an important function is the production and secretion of intrinsic factor, a compound that is necessary for effective absorption of vitamin B_{12} from the diet.

Q2 Peptic ulcers can occur in the duodenum and anywhere in the stomach, although they are usually located along the lesser curvature and in the pyloric region. Duodenal ulcers make up approximately 80% of the peptic ulcers diagnosed and occur in the first part of the duodenum.

Q3 The cells which produce HCl are the parietal cells; acid secretion can produce a stomach pH of 1.5–2. H^+ is secreted into the stomach lumen by an ATP-dependent proton pump in exchange for K^+. H^+ secretion depends on the dissociation of carbonic acid, formed by the hydrolysis of CO_2, in a reaction catalysed by carbonic anhydrase:

$$CO_2 + H_2O \xrightleftharpoons{\text{carbonic anhydrase}} H_2CO_3 \rightleftharpoons HCO_3^- + H^+.$$

As the H^+ is pumped into the lumen of the stomach, HCO_3^- moves out of parietal cells into blood and Cl^- enters the cell in exchange. Acid secretion is stimulated by histamine acting on H_2 receptors, by acetylcholine acting on muscarinic (M1) receptors and by gastrin acting on gastrin receptors of the parietal cells.

Pepsinogen is secreted by chief cells in the gastric mucosa and is the precursor of the protease enzyme pepsin.

In addition to HCl, the parietal cells produce intrinsic factor, which binds to dietary B_{12} and facilitates its absorption in the ileum. Gastrin, a hormone which promotes secretory activity in the stomach, is also produced by the gastric mucosal cells and released into the blood.

Q4 Water, mucus, pepsinogen and gastric lipase, which digests milk fats, are produced by the stomach. In young animals, rennin is also present in gastric secretions. The volume of secretion produced each day is approximately 2 l.

Q5 Gastric secretion is controlled in three phases:

(1) The cephalic phase, which is initiated by the sight, smell or thought of food before it enters the mouth. This is a nervous mechanism, mediated by the vagus.

(2) The gastric phase, which occurs when food actually enters the stomach. The presence of food and the composition of gastric contents stimulate local reflexes involving intrinsic nerve plexuses and stretch receptors and initiate release of gastrin from G cells to further enhance secretion.

(3) The intestinal phase, which occurs as chyme enters the duodenum. This involves many inhibitory controls: neural and endocrine mechanisms limit the rate of stomach emptying so that the secretory and absorptive mechanisms of the small intestine can cope effectively with the entry of gastric contents.

Q6 The stomach generates regular peristaltic waves which spread over the body of the stomach to the antrum and close the pyloric sphincter. Increased gastric motility occurs during a meal because of distension by the food, the activity of the vagus nerve and gastrin. When fat or excessive acid is present in the stomach, gastric emptying is slowed. Emptying chyme into the duodenum stimulates stretch receptors and initiates the enterogastric reflex, which temporarily inhibits gastrin secretion and gastric motility. The presence of chyme in the duodenum also causes release of secretin, CCK and gastric inhibitory peptide (GIP), which also reduce gastric activity. The main function of secretin and CCK is to stimulate release of fluid and enzymes from the pancreas to ensure that suitable pH and enzymic activity is available for digestion. The rate at which chyme enters the duodenum is increased if a meal is large, distends the stomach greatly and contains alcohol and/or wine. All these factors stimulate gastric secretion and motility.

Q7 Heartburn is a condition associated with reflux of gastric acid into the oesophagus. Normally, the contraction of circular muscle in the wall of the oesophagus acts like a lower oesophageal sphincter to prevent backflow (reflux) of acid material upwards from the stomach. When the stomach is full, lying down after a meal may allow acid to slip back along the oesophagus causing some patients to suffer burning pain in the chest (heartburn). When a person is upright, gravity helps to reduce the effects of acid reflux. The problem of acid reflux is particularly marked in patients with a hiatus hernia, a condition in which part of the stomach protrudes into the thoracic cavity. Such patients find that heartburn is made worse by bending or lying down. Sleeping propped up by pillows makes the upper body more vertical and allows gravity to assist in minimizing the effects of gastric reflux in affected individuals.

Part 2

Q8 A peptic ulcer is an area of erosion on the gastrointestinal mucosa. Ulcers commonly occur in the duodenum and stomach, causing discomfort and pain. Development of ulcers appears to be the result of an imbalance between the mucosal defence mechanisms, which include mucus production, and agents, such as acid and pepsin, which can erode the mucosa. If the quantity of acid increases or the efficiency of the defence mechanism diminishes, ulcers form.

Q9 The stomach secretes a very acid gastric juice with a pH of 1.5–2. The mucosa is normally protected from acid by a number of mechanisms. Mucus is produced by the large number of mucous cells in the body and fundus. It contains glycoproteins called *mucins*, and the mucus produced forms a kind of gel which coats the mucosal surface. In addition these cells secrete HCO_3^-, which is trapped in the mucus and increases the local pH to form a less acidic environment at the surface of the epithelial cells.

Prostaglandins (PGs) produced by the gastric mucosa stimulate the secretion of bicarbonate and mucus and inhibit the proton pump.

Agents which block the synthesis of PGs, for example non-steroidal anti-inflammatory drugs (NSAIDs) such as aspirin or ibuprofen, reduce the production of the protective PGs and predispose patients to the development of ulcers.

In recent years it has become clear that *H. pylori* infection is also important in the development of ulcers. The infection apparently upsets the balance between the protective mechanisms and the eroding effects of acid and pepsin: antibiotics are generally given to eradicate the infection and this is an effective treatment for achieving long-term healing of peptic ulcers.

Q10 Misoprostol is a synthetic PG analogue which is used to promote ulcer healing in patients who have ulceration related to use of NSAIDs, for example for chronic musculoskeletal pain. Its use can prevent the development of peptic ulcers as it has a protective effect on the mucosa.

Q11 Both alcohol and the caffeine content of coffee act directly on the gastric mucosa to stimulate acid and pepsinogen secretion, so reduction in their use should help the patient. Smoking can also worsen dyspepsia (indigestion) and heartburn, possibly via actions of nicotine on the stomach wall.

Q12 Pharmacological treatment of peptic ulcers aims to restore the balance between mucosal defence and mucosal damage by acid and pepsin in the stomach wall. The general mechanisms of drug action include: (i) inhibition of acid secretion, (ii) neutralization of the acid with antacid preparations, (iii) eradication of *H. pylori* with antibiotics and (iv) enhancement of the mechanisms which protect the mucosa.

Q13 Ranitidine is an example of an antagonist at histamine receptors on the parietal cells and has been in use for some years in the treatment of peptic ulceration. It blocks the H_2 receptor on these acid-secreting cells, so reducing or preventing the activation of the H^+–K^+ ATPase proton pump and the production of HCl. It can heal both gastric and duodenal ulcers.

Q14 Proton pump inhibitors, for example omeprazole, greatly reduce secretion of gastric acid by a direct inhibitory action on the proton pumps of the parietal cells, which secrete H^+ into the stomach lumen. They are used alone or in combination with the eradication of *H. pylori*, generally for treatment of severe reflux disease.

Q15 In addition to eradication of his *H. pylori* infection and a short course of a proton pump inhibitor, Patterson would benefit from regular small meals which include fruit and vegetables instead of one large meal at night. Preferably, he should have his last meal of the day several hours before retiring to bed. Reducing his consumption of alcohol and coffee, which stimulate acid secretion, and high-fat meals, which prolong the secretion of acid, will also help. Weight reduction, smoking cessation and, perhaps, raising the head of his bed a little should also help reduce his gastric reflux.

Key Points

- The stomach stores and mixes food with gastric acid secretions: stomach pH is 1.5–2. Gastric acid can erode areas of the stomach and duodenal mucosa to cause peptic ulcers if the alkaline mucus, which normally protects the mucosa, is reduced or lacking. Symptoms produced include: pain, nausea and vomiting. Alcohol, caffeine from coffee and smoking stimulate acid and pepsinogen secretion, making symptoms of peptic ulcer worse.
- Infection with the bacterium *H. pylori* is associated with ulcer development. Treatment of peptic ulcers may involve a course of antibiotics to eradicate this organism. Agents such as H_2 (histamine) antagonists, for example ranitidine, and proton pump inhibitors, such as omeprazole, are also used to heal peptic ulcers.
- PGs produced in the gastric mucosa stimulate the secretion of both bicarbonate and the mucus gel to protect the mucosa from damage by gastric secretions. Use of NSAIDs, such as aspirin and ibuprofen, reduces the production of the prostanoids and so decreases the protection of the gastric mucosa, promoting mucosal erosion.
- Gastric secretion and motility are controlled by both nervous and hormonal mechanisms. The vagus initiates the cephalic phase of secretion in response to the sight or smell of food, before food is eaten. The gastric phase occurs when food enters the stomach and is controlled both by intrinsic nerve reflexes in the stomach wall and released gastrin. The third, or intestinal, phase of secretion is coordinated by nervous and hormonal mechanisms to limit the release of the chyme from the stomach into the duodenum and reduces further acid secretion.

- Heartburn, a burning pain in the chest, is due to reflux of gastric acid from the stomach into the oesophagus. This worsens when patients lie down or bend down after eating a meal. If a person is in an upright position, gravity helps to reduce acid reflux. Heartburn is treated with proton pump inhibitors and H_2 antagonists. Patients are advised to reduce smoking, alcohol and coffee intake, to eat the last meal of the day some hours before retiring to bed and take regular, small, low-fat meals, rather than one large meal in the evening.

CASE STUDY 44 Daria's abdominal pain

Q1 The large intestine is approximately 1.2–1.5 m long and extends from the ileocaecal junction to the anus. Anatomically, the colon loops upwards, crosses the abdomen and descends to the rectum. It consists of four sections: the ascending, transverse, descending and sigmoid sections. In the muscular portion of the colonic wall, the outer, longitudinal muscle forms three bands called *taenia coli*. These bands of muscle are shorter than the length of the colon and give the colon a crinkled or gathered appearance. The circular muscles of the colonic wall separate the colon into pouches, or haustra.

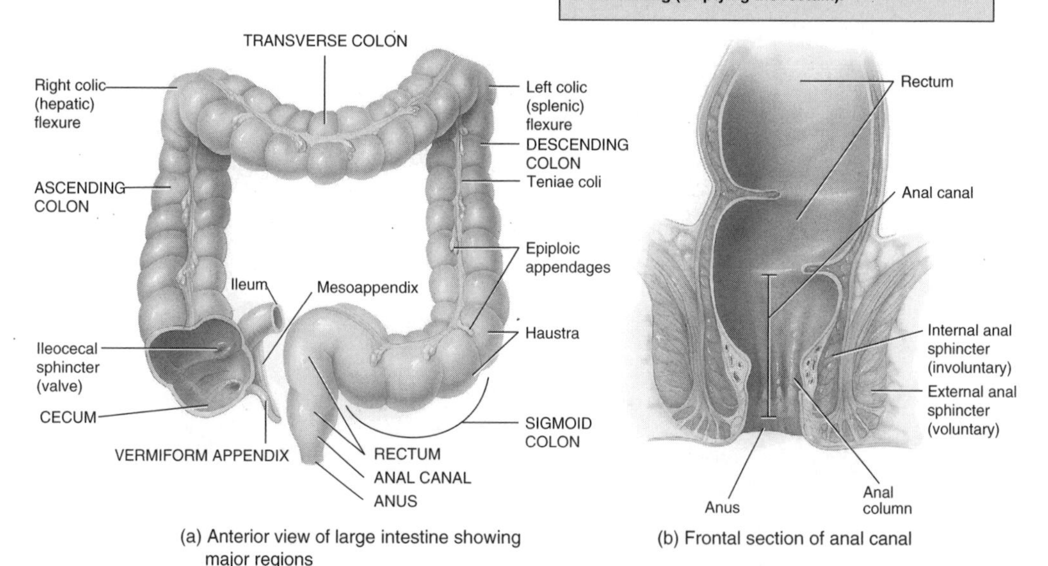

(a) Anterior view of large intestine showing major regions

(b) Frontal section of anal canal

From Tortora and Derrickson *Principles of Anatomy and Physiology, Eleventh Edition 2006.* Reproduced with permission of John Wiley & Sons, Inc.

The colon secretes an alkaline mucous which lubricates the passage of material through the large intestine. Approximately 500 ml of water and a considerable amount of sodium chloride enters the colon each day, and most is reabsorbed. Sodium is actively transported out of the colon lumen and water follows down an osmotic gradient. Approximately 100 ml of water remains in the faecal material to be excreted. No digestive enzymes are produced by the large intestine and no absorption of carbohydrate, protein or fat occurs in this part of the intestine.

Q2 Unlike the small intestinal mucosa, the colonic mucosa does not contain any villi. There are columnar epithelial cells and mucus-secreting goblet cells in the mucosa: the columnar epithelium reabsorbs fluid and electrolytes.

Q3 The muscular wall of the colon is quiescent for much of the time. The major type of motility is segmentation but this is less frequent than in the small intestine. It mixes and moves colonic contents around to promote reabsorption of water and electrolytes. Peristaltic movements also occur and promote colon emptying. The myenteric plexus in the wall of the colon coordinates both motor and secretory activity and affects activity of the internal sphincter at the junction of the colon and rectum. This sphincter is usually contracted and is maintained closed by sympathetic stimulation. Stimulation of parasympathetic nerves increases motor activity throughout the colon and relaxes the internal sphincter, allowing material to enter the rectum.

The gastrocolic reflex stimulates motility in the colon during or just after eating, when chyme enters the colon from the ileum. The content of the distal colon may be stored there for variable periods until defecation occurs, which may be 24 hours or more after eating. The rectum is normally empty and movement of faecal material into the sigmoid colon and rectum stimulates the defecation reflex.

Q4 An enormous number of bacteria are present in the lumen of the colon. About 95% of these are anaerobic and make up about one-third of the solid portion of the faeces; they play an important role in protecting the gut from exogenous infection.

The whole intestinal tract is sterile at birth but within three to four weeks the bacteria are established. These bacteria do not have significant digestive functions but are involved in metabolism of bile salts, sex steroids, some nitrogenous substances and drugs. The metabolism of colonic bacteria contributes to the production of gas (flatus), which is mostly nitrogen and carbon dioxide, but also contains methane and hydrogen sulphide. Some foods, for example beans, contain carbohydrate which cannot be digested in the small intestine; it is metabolized by colonic bacteria, producing large amounts of flatus. Some of the vitamin K which is required by the liver for the synthesis of a number of blood coagulation factors can be produced by bacteria in the colon, but the major source of this vitamin is dietary, in green, leafy vegetables.

Q5 Diverticula are sacs, or pouches, of the mucosa which form in the wall of the colon and bulge through the muscular wall of the intestine, often where arteries penetrate the intestinal wall. Their formation is associated with diets of low fibre content and is thought to be related to abnormal colon motility together with the development of high intraluminal pressure. These sacs may become infected and inflamed, causing pain. The presence of diverticula is known as *diverticulosis*; when infection and inflammation occur it is known as *diverticulitis*.

Q6 Diverticula are usually found in the sigmoid colon. The sigmoid colon is the section that follows the descending colon and forms the final part of the colon which leads into the rectum. It is located in the lower left side of the abdomen.

Q7 Fibre comes from the parts of plants which cannot be digested in the human gut. It may either be insoluble, for example cellulose, hemicellulose or lignins, or the soluble type, such as pectins and gums. The soluble type of fibre helps to slow absorption of cholesterol from the gut, lowers blood cholesterol and decreases the risk of coronary disease.
Insoluble fibre binds water, making the faeces softer and bulkier, aiding defecation.

Q8 Western diets contain a large proportion of refined foods and much less fibre than the diet of developing countries. Lack of dietary fibre reduces the volume of intestinal contents, making a smaller, harder faecal mass, which requires forceful contraction of the colon muscles and a high intraluminal pressure in the colon to move it along to the rectum. This appears to predispose patients to development of diverticula.
Addition of more insoluble fibre to the diet increases the bulk of the faeces, the rate of movement of gut contents and lowers intracolonic pressure. A normal transit time for food through the gut is approximately 20–48 hours, but many individuals have a transit time in excess of 72 hours.

Part 2

Q9 Daria's temperature was raised (38.5 °C) and her white cell count was higher than normal; both suggest that an infection was present.

Q10 The chance of developing diverticula increases with age and is more common in males than females. By the age of 50 approximately 20% to 50 % of people may have developed some diverticula, although the majority of people will have no symptoms. Two-thirds of the population over 85 years of age are thought to have diverticula, so the elderly population is most at risk, particularly if they eat a low-fibre diet and have a low fluid consumption.

Q11 Constipation is a common problem in Western society. The incidence is generally highest in the elderly and lowest in young people. Some causes of constipation in addition to low-fibre diets include: poor bowel habit–often the stimulus to empty the bowel is ignored; hormone disturbances, for example hypothyroidism and diabetes; intestinal nerve damage; immobility or lack of exercise; and medication, for example calcium channel blockers, iron supplements and opioids.

Q12 Irritation and damage to other structures in the abdomen may occur if the diverticulitis is not treated. Abdominal muscles may go into painful spasm and a minority of patients might have rectal bleeding. A major problem could be development of an intestinal obstruction or an abscess in the wall of the intestine. The abscess may eventually cause perforation of the intestinal wall; leakage of infected material into the peritoneal cavity and then infection of the peritoneal membranes (peritonitis) may occur. Peritonitis is a very serious condition.

Q13 Unless the diverticula are removed by a surgical procedure, which is usually *not* necessary, Daria has some risk of recurrence. However, if her diet is adjusted to contain more dietary fibre, transit of intestinal contents will be enhanced without the development of a high intraluminal pressure, and the chance of recurrence is greatly diminished. Fortunately, the majority of patients with known diverticulosis remain asymptomatic throughout their lifetime.

Q14 The recommended intake for dietary fibre ranges from 19 to 38 g per day. The average American consumes only approximately 14 g of fibre per day, so increasing fibre content to the recommended level is likely to require a considerable change in eating habits for many people. If the change to a high-fibre diet is made very quickly, diarrhoea and production of excess flatus is likely to occur. A further disadvantage of increasing fibre too much is that the absorption of minerals in the diet, such as iron, magnesium and zinc, may decrease. So good advice for this patient is: keep well hydrated and moderately increase consumption of fibre-rich foods, such as vegetables, nuts and whole-grain cereals. The proportion of these foods in her diet can then be gradually increased.

Key Points

- The colon and rectum form the large intestine and possess a mucosa composed of columnar epithelial cells and mucus-secreting goblet cells. The main function of the colon is to reabsorb water and ions, and the muscular wall shows less activity than the small intestine. No digestive enzymes are produced and no digestion or absorption occurs in this part of the intestine. The gastrocolic reflex stimulates peristaltic activity after eating when chyme enters the ileum. The colon normally stores gut contents for up to 24 hours. The rectum is usually empty: movement of faeces into the rectum stimulates the defecation reflex.

- The colon contains a vast number of bacteria which are involved in metabolism of bile salts, sex steroids and some drugs, and synthesis of vitamin K.
- Diverticula, which are pouches of the mucosa, can form and bulge through the muscular wall of the colon. These pouches may become infected and inflamed (diverticulitis), causing abdominal pain. Development of diverticula is associated with low-fibre diets and is thought to be related to abnormal colonic motility and high intraluminal pressures. The chance of developing diverticula increases with age.
- If diverticulitis is not treated, there may be irritation and damage to other abdominal structures. Intestinal obstruction or perforation of the intestinal wall could develop.
- Western diets are generally high in refined foods and low in fibre content. Fibre may be insoluble, for example cellulose, or soluble, for example pectin from plants. Insoluble fibre binds water, increases the bulk of the faeces and increases rapidity of transit through the intestine. Soluble fibre slows absorption of cholesterol and reduces blood cholesterol concentration.

CASE STUDY 45 That bloated feeling

Part 1

Q1 The mucosa is a mucous membrane which forms the innermost layer of the intestine. In the small intestine the mucosal surface area is increased greatly by folds and by villi, finger-like projections containing a core with a lymph capillary (lacteal) and blood vessels. Villi are covered by absorptive columnar epithelial cells whose luminal surface is further increased by microvilli (brush border) on which digestive enzymes and transport mechanisms for inorganic ions are located.

Between the villi are pits, the crypts of Lieberkühn, containing undifferentiated cells which move up to the tip of the villus, maturing in shape and function as they progress. They function for a few days and are then shed: the entire epithelial population is replaced every four to seven days.

The brush border of the villi hydrolyses oligosaccharides to glucose, fructose and galactose. Microvilli also attach peptidases, which hydrolyse di- and tripeptides to amino acids. The monosaccharides and amino acids are transported into the blood by active transport processes.

Q2 The final stage of carbohydrate and protein digestion takes place in the brush border of epithelial cells on the villi. In celiac disease the small intestinal mucosa becomes flattened because of inflammatory damage and atrophy of the villi. The resulting reduction in epithelial cells, containing the enzymes needed for the final digestion of disaccharides and peptides, greatly reduces digestion of carbohydrate and protein and significantly diminishes the absorption of these nutrients.

Q3 Malabsorption interferes with absorption of nutrients. Patients may suffer: weight loss, easy fatigability and muscle weakness, tiredness, diarrhoea, flatulence, abdominal distension and anaemia.

Q4 The water-soluble vitamins (B and C) and essential minerals, such as iron, are absorbed in the small intestine. The reabsorption of iron in the duodenum and proximal jejunum involves a complex active transport process. When there is a large reduction in the surface area of this part of the gut, there is a marked reduction in reabsorption of iron (and B vitamins). Haemoglobin synthesis is decreased, leading to development of anaemia, which is a common symptom in celiac disease.

Q5 Chloe appears to be suffering from iron-deficiency anaemia as her red blood cells are small (microcytic). In addition her ferritin level is low. Iron is stored as ferritin, so a reduction in ferritin concentration reflects a decreased store of iron in the body, a state which is characteristic of iron-deficiency anaemia. In other forms of anaemia the iron stores are normal or may be raised. Chloe's vitamin B_{12} concentration, although low, is within normal limits, so she is unlikely to be suffering from pernicious anaemia.

Q6 Vitamin B_{12} is absorbed from the terminal ileum. For successful absorption of this vitamin, intrinsic factor from the stomach is required. Since the stomach is not affected by celiac disease, production of intrinsic factor is not reduced. The terminal ileum is usually little affected by celiac disease, perhaps because the toxic components of gluten have been digested or inactivated in some way before the intestinal contents reach this part of the intestine. The absorption and blood concentration of vitamin B_{12} in celiac patients is usually within normal limits.

Q7 Failure to absorb adequate nutrients from ingested food leads to symptoms of malnutrition, including weight loss, weakness, fatigue, anaemia and reduction of bone density. If Chloe's mucosal surface has been reduced to approximately half, her ability to absorb basic nutrients, including dietary iron needed for haemoglobin synthesis, is similarly reduced. Reduced haemoglobin synthesis results in anaemia and a decreased oxygen-carrying capacity of blood, resulting in reduced exercise capacity, weakness and extreme tiredness.

Part 2

Q8 Celiac patients are intolerant of gluten, a protein containing the peptide alpha-gliadin (α-gliadin), which appears to be toxic and injures their mucosal cells. The mechanism is possibly immunogenetic, but viral damage may also be involved. There is a higher frequency of celiac disease in patients of Northern European, English and Irish origin than in Asian or African patients.

Q9 Lipid in the diet is present mostly in the form of triglycerides, which are digested by pancreatic lipase to yield fatty acids and monoglycerides: bile salts are also required for digestion and absorption of the dietary lipids. Bile salts interact with the fatty acids and monoglycerides in the gut lumen to form micelles, which can be absorbed by the epithelial cells. In the epithelial cell the triglyceride is resynthesized to form droplets, or chylomicrons, which enter the lacteals and are carried by the lymphatic system into the general circulation.

Q10 Loss of the villi and epithelial cells causes failure of absorption of fats and their digestive products from the intestinal lumen, although lipase and bile salt production may be adequate. Large quantities of fat can remain in the intestinal contents to be excreted in the faeces: this is steatorrhoea. Faeces with a high fat content float in water and are difficult to flush away.

Q11 The major problem in celiac disease is loss of a substantial proportion of the absorptive surface of the small intestine. Because carbohydrate, protein and fat are not being completely digested or reabsorbed, a large amount of osmotically active material with retained fluid passes on to the colon. In the colon, bacteria metabolize the material, producing more osmotically active particles and further fluid retention, which accounts for the diarrhoea experienced by celiac patients. Another result of bacterial action in the colon is the production of gasses, mainly CO_2 and H_2. This produces flatulence, adds to the bloated feeling and causes the 'explosive' expulsion of faeces.

Q12 When the intestinal mucosa was damaged by gluten in Chloe's diet, it was not possible to satisfactorily absorb iron. While the damage persisted, it is unlikely that much extra absorption could occur when additional iron is ingested. By excluding gluten from the diet, the intestinal mucosa is allowed to recover and approach normality. At this point at least some of the additional iron given therapeutically is likely to be absorbed from the gut. So iron supplements are likely to be effective at this point.

Q13 Osteomalacia and osteoporosis are complications of celiac disease. The mineral in bone is mainly calcium phosphate; a supply of calcium is therefore needed for bone growth and replacement. Calcium is absorbed by active mechanisms in the duodenum and jejunum, facilitated by a metabolite of vitamin D. It is also passively absorbed in the ileum and specific calcium binding proteins are present in the intestinal epithelial cells. Loss of absorptive cells and calcium binding proteins markedly decreases calcium uptake and limits its availability for bone growth and repair.

Owing to decreased uptake of dietary calcium, celiac patients are at more risk of reduced bone mass (osteoporosis) and subsequent fracture than healthy individuals. This is particularly the case in female patients who have been diagnosed as adults. All women experience some increased bone loss following the menopause, because of decline in oestrogen production: bone loss is greater when absorption of calcium is also compromised. Bone density peaks at around 27–35 years of age, and after the age of 40 both males and females begin to lose bone mass. The rate of calcium loss increases in females following the menopause and it is therefore important that bone density is optimized in the years when bone density is increasing or peaking. After this point bone loss is inevitable and, if osteoporosis occurs, bone fractures are more likely to occur and can lead to considerable disability.

Q14 It is possible but unlikely. There is evidence of a genetic predisposition to celiac disease and the onset of symptoms can be triggered in susceptible individuals by an environmental event. The amount of gluten that can be taken by a susceptible individual, without provoking symptoms, varies greatly: some patients are much more sensitive than others and the immune reactivity of the mucosa also varies with time and hormonal status. Ingestion of gluten is more likely to cause a relapse if the intestinal wall is already inflamed, for example following a viral illness. So while some patients may be able to eat small amounts of products containing gluten with no ill effects, others suffer relapse when gluten is ingested.

Complete avoidance of gluten is difficult as it is found in a range of foods, including confectionary and manufactured products, which appear unconnected with wheat or other cereals. Celiac disease is a life-long condition which cannot be cured, but it can be successfully managed by following a gluten-free diet.

Key Points

- The mucosa of the small intestine has an enormous surface area because of the presence of villi. Villi are covered by absorptive columnar epithelial cells whose surface is further increased by microvilli (brush border), on which carbohydrate and peptide digestive enzymes and transport processes involved in absorption are situated. Pits between the villi contain undifferentiated cells which move up the villi, mature, function for a few days and are shed into the lumen of the gut.
- Pancreatic amylase hydrolyses starches to maltose and oligosaccharides. Final digestion of carbohydrates takes place in the brush border of the epithelial cells. Oligosaccharides are hydrolysed to monosaccharides and reabsorbed. Proteins are hydrolysed to peptides by pepsin, trypsin and several other proteolytic enzymes from the pancreas. The brush border attaches proteases which hydrolyse the di- and tripeptides to amino acids and absorbs them.
- Celiac patients have an allergy to gluten in wheat, which is toxic to epithelial cells on the villi; epithelial cells are destroyed and the small intestinal mucosa becomes flattened. The absorptive surface of their intestine is therefore greatly reduced, leading to malabsorption of nutrients and, in children, reduced growth. Failure to absorb adequate nutrients from the diet results in weakness, weight loss and reduction in bone density. Since the dietary components are not absorbed, they remain in the gut to exert osmotic effects and fluid retention. Bacteria in the colon metabolize the nutrients creating gas and further osmotically active materials: these products cause bloating and episodes of explosive diarrhoea.
- Lipids in the diet are also poorly absorbed following the loss of villi, and a large proportion of undigested fat remains in the faeces, which float and are difficult to flush away.

- Water-soluble vitamins and minerals such as iron are absorbed in the small intestine. Since celiac patients have a reduced absorptive area, absorption of iron and vitamins is reduced and anaemia is common.
- Celiac patients must avoid gluten in order to restore their absorptive surface towards normal. For most patients this involves a lifetime of careful scrutiny of the composition of manufactured foods and elimination of many popular commercial products from their diet.

9
Autonomic disorders

CASE STUDY 46 Rob's ocular accident

Part 1

Q1 The iris contains pigment cells, blood vessels and two layers of smooth muscle fibres. The papillary constrictor muscles are arranged in concentric circles round the pupil; when they contract, the size of the pupil decreases. The papillary dilator muscles are arranged radially away from the edge of the pupil. Contraction of these muscles widens, or dilates, the pupil. Both sympathetic and parasympathetic neurones supply and control the function of the muscles in the iris.
The stimulation of sympathetic neurones contracts the radial muscles and so dilates the pupil, whilst parasympathetic stimulation causes contraction of the concentric sphincter muscle, so constricting the pupil.

Q2 *Mydriasis* means the dilation of the pupil. *Miosis* is the constriction of the pupil. Muscarinic antagonists and alpha-adrenoceptor (α-adrenoceptor) agonists cause dilation of the pupil. However, muscarinic agonists and α-adrenoceptor antagonists cause contraction of the pupil.

Q3 The causes of unequal size of the pupils could be:

(1) Inflammation of the anterior chamber, which causes spasm of the sphincter muscle

(2) Acute closed-angle glaucoma

Clinical Physiology and Pharmacology Farideh Javid and Janice McCurrie

(3) Congenital malformation of the iris

(4) Exposure to drugs which affect the autonomic control of the pupils.

Patients with unequal size of the pupils because of local causes have a painful red eye with a small pupil and suffer visual disturbances (photophobia).

Q4 Yes. Rob has used both atropine and phenylephrine this afternoon. Muscarinic antagonists such as atropine, tropicamide and cyclopentolate cause dilation of the pupils. The α-adrenoceptor agonists, such as phenylephrine, also produce my driasis. Mydriasis may cause acute closed-angle glaucoma in some patients. It is unlikely that a very small amount of cocaine in the eye would cause problems, but in cocaine overdose pupils become widely dilated. This is due to blockade of uptake 1, a process normally involved in terminating the effects of noradrenaline. In the presence of cocaine the effects of sympathetic stimulation on the eye would be prolonged and the pupil would dilate. Morphine causes constriction of the pupils via opiate receptors.

Part 2

Q5 The doctor/ophthalmologist should lower the pressure in Rob's eye as quickly as possible to prevent damage.

Q6 Pilocarpine eyedrops are suitable. In severe conditions, in addition to the eyedrops, intravenous acetazolamide and intravenous hypertonic mannitol (an osmotic agent) may be used to reduce pressure. Acetazolamide prevents the actions of carbonic anhydrase in the ciliary body and inhibits bicarbonate synthesis. This causes reduction in sodium transport and aqueous humour formation since there is a link between bicarbonate and sodium transport.

Q7 *Glaucoma* is a term which describes a group of ocular diseases involving a loss of visual field and alteration in the optic disc. These conditions usually develop when there is an abnormally high intraocular pressure (IOP) and can lead to optic nerve damage and loss of vision if not treated. Glaucoma is due to the prevention of the outflow of the aqueous humour from the anterior chamber of the eye. The optic disc becomes concave because of the damage to the nerve fibres. There are three types of glaucoma: congenital glaucoma, closed-angle glaucoma and open-angle glaucoma.

The ciliary epithelium in the posterior chamber secretes aqueous humour. The aqueous humour flows in between the cornea and iris. After filtration through the trabecular meshwork, it returns to the venous circulation via the canal of Schlemm.

Patients with open-angle glaucoma develop an abnormal increase in IOP, which is due to an abnormality of the trabecular meshwork. This can happen in the absence of an obstruction between the trabecular meshwork and the anterior chamber. This condition is chronic and progressive and often has no symptoms. If it is not treated, the condition can result in loss of visual field. Corticosteroids can increase the IOP in some patients.

Patients with closed-angle glaucoma have a small angle between the iris and cornea which can also completely close and therefore flow of aqueous humour will be prevented. This results in a rise in IOP, which must be reduced quickly to prevent any damage to the retina.

In congenital glaucoma the trabecular network is attached to the iris. Blindness occurs if surgery is not performed to correct this problem.

Q8 Normal IOP is 15 mmHg (range is 10–20 mmHg). It is maintained by a balance between the secretion of aqueous humour by the ciliary body and its flow into the canal of Schlemm via the trabecular meshwork. Glaucoma is present when IOP rises to >21 mmHg.

Q9 Different classes of drugs can be used:

(1) beta-adrenergic (β-adrenergic) antagonists, such as timolol, betaxolol hydrochloride, carteolol hydrochloride, levobunolol hydrochloride, metipranolol. They cause a reduction in the secretion of aqueous humour by the ciliary body which in turn lowers IOP. This is done by antagonizing beta-2-adrenoceptors (β_2-adrenoceptors) on the ciliary body and therefore reducing aqueous secretion. These drugs may also cause vasoconstriction of the vessels supplying the ciliary body, which then leads to reduction in aqueous secretion.

(2) Sympathomimetic agents, such as brimonidine tartrate, apraclonidine, adrenaline and dipivefrine hydrochloride, which is a prodrug for adrenaline), act on α-adrenoreceptors to induce dilation of the veins to reduce IOP. They also induce mydriasis (dilation of the pupils). Adrenaline may reduce the rate of formation of the aqueous humour, which in turn reduces the IOP; it may also increase the outflow through the trabecular meshwork. Stimulation of alpha-2-adrenoreceptor (α_2-adrenoreceptor) by drugs such as brimonidine and apraclonidine on the adrenergic neurons supplying the ciliary body can also result in reduction of secretion of aqueous humour.

(3) Parasympathomimetics such as physostigmine sulfate increase the level of acetylcholine, which contracts the ciliary muscle and opens the fluid pathway which leads to reduction in IOP. They are usually used with pilocarpine hydrochloride or pilocarpine nitrate, which acts on cholinoceptors. Parasympathomimetics cause poor night vision because of miosis.

(4) Carbonic anhydrase inhibitors, such as acetazolamide and dichlorphenamide, act as diuretics to increase excretion of water by inhibiting carbonic anhydrase activity. This in turn leads to a reduction in the level of bicarbonate in aqueous humour.

(5) Prostaglandin-related drugs, such as latanoprost, which is a prodrug of prostaglandin-F_2, reduce IOP by passing through the cornea and increasing the outflow of aqueous humour.

Q10 Beta-adrenoceptor antagonists are contraindicated in patients with asthma or respiratory obstructive diseases, bradycardia, heart block or heart failure. Adrenergic agonists are contraindicated in patients with closed-angle glaucoma and should be used cautiously in patients with hypertension or heart disease. Parasympathomimetics cause poor night vision and dimming of vision, because of development of miosis, headache and brow ache. Carbonic anhydrase inhibitors have a weak diuretic action and can induce depression, drowsiness, paraesthesia, electrolyte disturbance such as hypokalaemia, acidosis and lack of appetite.

Q11 Yes. Laser surgery can be used to form a hole in the iris to permit increased flow of aqueous humour.

Key Points

- Mydriasis (dilation of the pupil) can be caused by muscarinic antagonists and α-adrenoceptor agonists. Mydriasis may cause acute closed-angle glaucoma in some patients.
- Miosis (constriction of the pupil) can be caused by muscarinic agonists and α-adrenoceptor antagonists.
- The causes of unequal size of the pupils include:
  - inflammation of the anterior chamber, causing spasm of the sphincter muscle
  - acute closed-angle glaucoma
  - congenital malformation of the iris
  - exposure to drugs which affect the autonomic control of the pupils.
- Glaucoma is a group of ocular diseases involving a loss of visual field and alteration in the optic disc. Glaucoma usually develops in response to high IOP, leading to optic nerve damage and loss of vision.
- There are three types of glaucoma: congenital glaucoma, closed-angle glaucoma and open-angle glaucoma.
- Drug treatment involves use of β-adrenergic antagonists, sympathomimetic agents, parasympathomimetics in conjunction with pilocarpine hydrochloride or pilocarpine, carbonic anhydrase inhibitors or prostaglandin-related drugs.

CASE STUDY 47 A severe attack of greenfly

Part 1

Q1 The insecticide can enter the body via the lung by inhalation and, if splashed onto bare skin, can be absorbed via the skin, mucous membranes and eyes. Possibly, Jim also swallowed some of the spray.

Q2 Both the central and peripheral parts of the nervous systems have been affected by the insecticide, which has produced effects mediated by both muscarinic and nicotinic receptors.

Q3 The pathway followed by autonomic nerves involves a chain of two neurones. The cell body of the first neurone in the chain lies within the central nervous system (CNS) but its axon, the preganglionic fibre, synpases outside the CNS. In the sympathetic nervous system the preganglionic nerves originate in the thoracic and lumbar regions of the spinal cord. Most of the preganglionic neurones are short and most synapse with cell bodies of postganglionic neurones in a chain of ganglia lying close to the spinal cord (paravertebral ganglia). Some sympathetic preganglionic fibres synapse in collateral ganglia: the celiac, superior mesenteric and inferior mesenteric ganglia. Postganglionic nerves of the sympathetic system are generally long.
In the parasympathetic system the preganglionic fibres originate in the brain and in sacral regions of the spinal cord. These fibres are long compared to those in the sympathetic system and their synapse with the postganglionic neurone usually lies in the tissues innervated. The vagus nerve, which arises from the cranial outflow, is a major component of the parasympathetic system and provides innervation to the heart, lungs, oesophagus, stomach, small intestine and upper large intestine. The innervation supplying the lower parts of the large intestine, reproductive tissues and urinary system arises from sacral areas of the cord.

Q4 The ganglionic transmitter of both divisions of the autonomic nervous system is acetylcholine. The major postganglionic neurotransmitter of the sympathetic nervous system is norepinephrine (noradrenaline), but a small number of structures are innervated by sympathetic, cholinergic fibres. These fibres release acetylcholine and the structures innervated include the sweat glands and blood vessels supplying skeletal muscle. In the parasympathetic system the postganglionic neurotransmitter is acetylcholine.

Q5 The parasympathetic transmitter acetylcholine is stored in synaptic vesicles within the postganglionic nerve ending. When an action potential arrives at the parasympathetic nerve terminal, the nerve ending depolarizes and voltage-gated calcium channels open. Calcium diffuses into the nerve terminal down an electrochemical gradient from the extracellular fluid (ECF). The increase in intracellular calcium within the terminal stimulates synaptic vesicles to fuse with the presynaptic membrane and acetylcholine is released into the synaptic cleft. Acetylcholine diffuses across the cleft to the postsynaptic membrane and binds with muscarinic receptors; this leads to opening of Na^+ channels and propagation of the action potentials in the postsynaptic area. The remaining acetylcholine will be metabolized by cholinesterases and, as a result, choline will be taken back up to the presynaptic area.

Q6 In comparison with the sympathetic transmitter norepinephrine, the inactivation of acetylcholine by cholinesterases is rapid so that normally the activity of acetylcholine at the synapse is relatively short-lived. The choline component is taken up into the presynaptic terminal and acetylcholine is resynthesized and stored in the synaptic vesicles. Anticholinesterases function as cholinergic stimulants in the parasympathetic nervous system since they greatly prolong and so increase the actions of endogenous acetylcholine at muscarinic receptors on the effector tissue.

Q7 Intestinal cramps are produced by the intense stimulation of muscarinic receptors (mainly type M3) on intestinal smooth muscle, because of prolonged activity of released acetylcholine from postganglionic parasympathetic nerves. Intestinal motility and secretion are greatly stimulated while the sphincters are relaxed, leading to rapid transit of the gut contents and diarrhoea. Lacrimal glands are similarly stimulated by activation of tissues innervated by the parasympathetic nervous system, causing eyes to water profusely (lacrimation).

Sweating is stimulated by direct action of the increased endogenous acetylcholine on muscarinic receptors of sweat glands in the skin: these glands are innervated by the sympathetic division of the autonomic nervous system.

Agitation is produced by a central excitatory effect on cholinergic neurones in the brain, but in large doses anticholinesterases can cause depression of the respiratory centre in the medulla.

Skeletal muscle twitching is due to effects at the skeletal neuromuscular junction, which is innervated by the somatic nervous system, via motor nerves. The anticholinesterase prolongs and intensifies the actions of released acetylcholine at the junction, causing fasciculation (strong, jerky contractions) of skeletal muscle. Normally at the skeletal neuromuscular junction, the released acetylcholine is rapidly hydrolysed by cholinesterases to choline and acetate. This allows repolarization of the muscle membrane to occur following initial stimulation. In the presence of anticholinesterases the acetylcholine remains at the junction for a very prolonged period and produces repeated twitching of the muscle fibres via nicotinic receptors.

Part 2

Q8

(1) The prolonged action of acetylcholine at the parasympathetic nerve ending would greatly slow the rate of the heart (bradycardia) and also slow conduction of the cardiac impulse over the atria and the atrioventricular (AV) node.

(2) In the bronchi, prolonged survival of acetylcholine would cause intense bronchoconstriction and increased secretion of mucus with resulting dyspnoea.

(3) The salivary glands are innervated by parasympathetic nerves. Increased activity of acetylcholine in this location would greatly increase the secretion of saliva in the mouth and would account for Jim's drooling.

Q9 The cholinergic effect of the anticholinesterase, operating via muscarinic receptors, causes intense bronchoconstriction and a considerable increase in fluid secretion into the bronchial lumen. This would cause Jim some obstructive problems with his breathing. If exposure to the malathion was prolonged, the respiratory centre in the medulla is likely to be depressed, so that support to his breathing would be required for a time, until the effects of the chemical had diminished. Assisted ventilation, frequent removal of bronchial secretions and oxygen are likely to be required.
The effects of the prolonged action of acetylcholine on skeletal muscle may also reduce the respiratory efficiency of the diaphragm and chest wall muscles, which would limit ventilation of Jim's lungs.

Q10 Atropine would be a suitable antagonist to the effects of the anticholinesterase at muscarinic receptors and would reverse the intestinal cramps, lacrimation, drooling, sweating and so on, since atropine is a muscarinic antagonist. So atropine can be considered as an 'antidote'.

Q11 No. The receptors at the neuromuscular junction are nicotinic and would not be affected by atropine, which is a muscarinic antagonist.

Q12 Bethanechol is a parasympathomimetic agent which mimics the actions of acetylcholine. It is capable of stimulating the muscarinic receptors on the detrusor muscle and can contract the bladder, leading to more effective emptying of urine.
Bethanechol does, however, have some side effects: it can cause nausea, vomiting, intestinal colic, blurred vision, sweating and bradycardia if given in an excessive dosage.

Part 3

Q13 Agents used in eye examinations are intended to relax the sphincter muscle of the iris, dilating the pupil and allowing visualization of the retina: they are known as *mydriatics*. These agents are muscarinic antagonists like atropine, although the agent usually preferred is tropicamide as it is a shorter-acting anticholinergic drug. Atropine is a long-acting mydriatic and has had many different clinical uses in the past. It causes cardiac quickening via an action on the pacemaker tissue of the heart and dries the oral mucosa by reducing the secretion of saliva.

If atropine is administered in large amounts, it can cause a number of systemic effects, for example relaxing muscular spasm in the gastrointestinal tract, gall bladder, urinary bladder and ureter. It has been used to reverse bradycardia and dry the secretions of bronchial and oral mucosa prior to surgery.

When taken in very large amounts, atropine poisoning occurs; this may, for example, be seen following poisoning by deadly nightshade berries. These berries can cause symptoms which patients have colourfully described as:

- dry as a bone: decreased salivary secretion
- hot as a furnace: inhibition of sweating
- blind as a bat: dilation of the pupil and paralysis of the ciliary muscle, which prevents accommodation
- mad as a hatter: stimulatory effects in the CNS.

Key Points

- The pathway followed by autonomic nerves involves a chain of two neurones. The ganglionic transmitter in both sympathetic and parasympathetic ganglia is acetylcholine. In the sympathetic system the preganglionic nerves emerge from thoracic and lumbar regions of the spinal cord, most are short and synapse in ganglia lying close to the spinal cord. The postganglionic nerves are long and the postganglionic transmitter is either noradrenaline or acetylcholine. In the parasympathetic system preganglionic nerves originate either in the brain or in sacral regions of the spinal cord. The nerve is usually long and synapses in ganglia within the tissues innervated. The postganglionic nerve is normally short and the transmitter is acetylcholine.
- The activity of acetylcholine at synapses is usually short-lived. Released acetylcholine is hydrolysed by cholinesterases. The choline component is taken up into nerve terminals and used to re-synthesize the transmitter. Anticholinesterases inhibit the activity of cholinesterase enzymes and allow

acetylcholine to persist at the synapse, increasing its action at muscarinic and nicotinic receptors.

- Anticholinesterases such as malathion are used in commercial insecticide sprays. Unprotected operators may absorb malathion via the eyes, skin, respiratory tract and mucous membranes of the mouth. Effects include: intestinal cramps and diarrhoea following stimulation of intestinal motility and secretion. Stimulation of lacrimal and salivary glands causes the eyes to water profusely (lacrimation) and saliva to drool. Bradycardia, bronchoconstriction, dyspnoea and increased sweating also occur. Skeletal muscle twitching (fasciculation) is due to the prolonged action of released acetylcholine at the skeletal neuromuscular junction.
- The central actions of the anticholinesterase cause agitation because of a prolonged excitatory effect of released acetylcholine on cholinergic neurones in the brain, but in large doses anticholinesterases can cause depression of the respiratory centre in the medulla.
- A suitable treatment to reduce many of the effects of the anticholinesterase would be atropine, a muscarinic antagonist. This would not reverse the twitching of skeletal muscle, which is due to stimulation of nicotinic receptors. Cholinesterase activity in the body can be reactivated by the drug pralidoxime, which must be given soon after exposure to the anticholinesterase.
- Atropine is used in eye examinations to dilate the pupil and allow visualization of the retina.

10

Reproductive disorders

CASE STUDY 48 Panic of a college girl

Q1 Menstruation is a process in which the endometrial lining of the uterus in a non-pregnant woman is shed over a period of about three to five days, usually every 28 days, following a change in the level of ovarian hormones. According to the changes in the ovary and the endometrium, the menstrual cycle is divided into: (i) the follicular phase and (ii) the luteal phase. From the first day of menstruation up to the day of ovulation, the ovaries are in the follicular phase. The luteal phase starts after ovulation, lasting until the first day of menstruation. The cyclical changes that occur in the endometrium are the menstrual, proliferative and secretory phases.

Q2 On day 1 of the cycle, increase in the level of gonadotropic hormones triggers the release of follicle stimulating hormone (FSH) and luteinizing hormone (LH). FSH and LH stimulate growth of the follicle and its maturation, and also secretion of oestrogens. When the level of oestrogens increases in the blood, negative feedback inhibits the anterior pituitary gland, reducing the release of (mainly) FSH and LH.

High levels of oestrogens then induce a positive feedback on the anterior pituitary in the late follicular phase to cause an increase in the level of LH. Following the LH surge, ovulation occurs and the ruptured follicle is converted to a corpus luteum.

The corpus luteum causes an increase in the levels of oestrogens and progesterone which in turn induces negative feedback to inhibit LH and FSH release.

Clinical Physiology and Pharmacology Farideh Javid and Janice McCurrie

If fertilization does not occur, the level of LH will be reduced and luteal activity ends; the corpus luteum then degenerates, resulting in reduced levels of oestrogen and progesterone which triggers menstruation. The cycle then restarts.

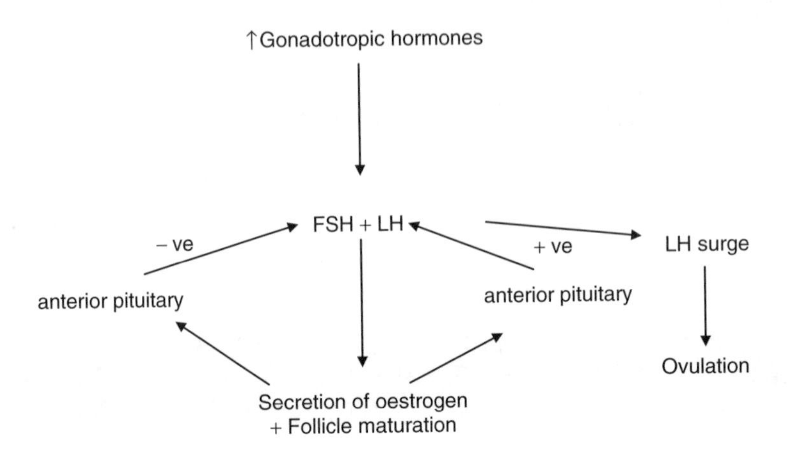

Q3 The absence of menstruation is called *amenorrhoea*. In primary amenorrhoea, a female has not reached menarche (the onset of puberty) by the age of 16 years. This might be a delayed puberty due to high levels of exercise with reduction in body weight and fat, or to congenital defects in the production of gonadotrophin which prevents the development of the gonads. In primary hypogonadism, for example Turner's syndrome, which is caused by possession of only one X chromosome, the body form is female but the ovaries are non-functional and lack of oestrogen accounts for the amenorrhoea.

Secondary amenorrhoea is defined as absence of menstruation for three or more cycles in a female who has previously menstruated. Polycystic ovarian syndrome is a common cause of amenorrhoea in young women. Amenorrhoea may also follow rapid loss of body weight when dieting and is associated with anorexia nervosa. Pregnancy must always be considered when amenorrhoea occurs, but many other conditions such as severe illness or stress and stopping the contraceptive pill may also result in cessation of periods.

In terms of pharmacological management, often little can be done for those with an abnormality in the hypothalamic–pituitary–gonad axis. In some patients hormone replacement is required, and in patients with Turner's syndrome *in vitro* fertilization and hormone therapy can be offered.

Q4 Excessive bleeding during the menstrual period is called *menorrhagia*. The blood loss reduces levels of iron in the body and may result in iron-deficiency anaemia. The causes of excessive bleeding could be inflammation, fibroids, endometriosis, cervical polyps, adenomyosis, ovarian tumours, intrauterine devices (IUDs), inherited clotting disorders, endocrine dysfunction, such as thyroid dysfunction, or mental stress. In terms of drug therapy, oral ferrous

sulphate (200–600 mg daily in divided doses) can restore the iron stores; treatment needs to be given for several months. Tranexamic acid, 1 g three times daily for four days, inhibits fibrinolysis, promotes the coagulation of blood and so reduces blood loss in patients with menorrhagia.
In some patients endometrial proliferation may be decreased by treatment with an intrauterine progestogen, for example levonorgestrel: within three to six months menorrhagia can be considerably reduced.

Q5 There are basically three categories of contraceptive methods:

(1) Hormonal contraceptives, which use oral oestrogen and/or progesterone, such as combined hormonal contraceptives and progestogen-only contraceptives.

(2) Contraceptive devices, for example IUDs. These usually consist of a plastic T-shaped frame fitted with copper bands or wires. There are also intrauterine progesterone-only devices, which release levonorgestrel directly into the uterus.

(3) Use of barriers, such as condoms, diaphragms and caps used with spermicides. Other contraceptives are: postcoital oral contraceptives (which can be taken up to 72 hours after unprotected intercourse), depot progestogen (which is given intramuscularly), progestogen implants, vaginal rings impregnated with hormone, contraceptives for males and vaccines (under research).

Q6 There are two main types of oral contraceptive:

(1) The combined pill, which consists of both oestrogen and progestogen, is taken for 21 consecutive days followed by seven pill-free days. The oestrogen prevents the release of FSH, which results in suppression of the follicle in the ovary. The progestogen prevents the release of LH, so preventing ovulation; it affects the cervix and endometrium to decrease sperm viability. Both hormones act to change the endometrium to discourage implantation and interfere with the coordinated contractions of the cervix, uterus and Fallopian tubes, which are essential for fertilization and implantation.

(2) The progestogen-only pill, which contains progestogen alone. The pill is taken daily without interruption. It makes the cervical mucus inhospitable to sperm. It may also change the endometrium to discourage implantation, and modify the coordinated contractions of the Fallopian tubes.

Q7 The combined pills are a safe and effective method of contraception for most women. They decrease the incidence of: amenorrhoea, irregular periods and intermenstrual bleeding, iron-deficiency anaemia, premenstrual tension and the incidence of benign breast disease, uterine fibroids and functional cysts of the ovaries. However, the associated side effects in some individuals might be: weight gain, nausea, flushing, dizziness, depression or irritability and amenorrhoea of variable duration following withdrawal. Use of combined oral contraceptives may increase risk of cardiovascular disease and impair glucose tolerance in those predisposed to diabetes mellitus.

Following use of oral contraceptives, females normally retain fertility, and the normal cycles of menstruation usually start soon after withdrawing the pill. There is evidence both for and against an increased risk of breast cancer.

The use of the progestogen-only pill is less reliable than the combined pill, and irregular bleeding might occur.

Q8 Human chorionic gonadotrophin, which is produced by the implanted blastocyte (developed from the fertilized egg) in the uterus, maintains the corpus luteum and its production of oestrogen and progesterone. This inhibits menstruation; therefore implantation of the embryo can occur, and subsequently a placenta can be formed.

Q9 The hormone that contracts the uterus during labour is the peptide hormone oxytocin, which is released from the posterior pituitary gland.

Q10 Missing a period, while establishing the normal menstrual cycle, is not unusual in a young adolescent girl. In older adolescents like Jane, her weight loss and/or the stress of her studies and the forthcoming examinations may have played an important role in delaying or disrupting her menstrual cycle.

Key Points

- Menstruation involves shedding the endometrial lining of the uterus over a period of about three to five days, usually every 28 days, following changes in the level of ovarian hormones. The absence of menstruation is called *amenorrhoea* and excessive bleeding during the menstrual period is called *menorrhagia*. The latter can be treated with tranexamic acid, which inhibits fibrinolysis and promotes blood coagulation.
- According to the changes in the ovary and the endometrium, the menstrual cycle is divided into: (i) the follicular phase and (ii) the luteal phase.
- Gonadotrophic hormones trigger the release of FSH and LH.
- FSH and LH stimulate the growth of the follicle and its maturation, and also secretion of oestrogens.

- Human chorionic gonadotropin maintains the corpus luteum and its production of oestrogen (estradiol) and progesterone. This inhibits menstruation, allowing implantation of the embryo to occur, and subsequently a placenta is formed. Contraceptive methods include oral contraceptives, contraceptive devices, for example IUDs, use of barriers (such as condoms, diaphragms and caps), postcoital oral contraceptives, depot progestogen, progestogen implants and vaginal rings impregnated with hormone.
- The combined oral contraceptive pill contains both oestrogen and progestogen.
- The progestogen-only contraceptive pill contains progestogen alone.

CASE STUDY 49 Shabana's monthly problems

Part 1

Q1 From the symptoms of pain, nausea and headache which Shabana described, and which began around the time that bleeding occurred each month, she appears to have suffered from dysmenorrhoea. Since Shabana also experienced changes in mood, some form of premenstrual syndrome (or premenstrual tension) might also have been present.

Q2 The symptoms associated with *premenstrual syndrome* are very varied. It is a term used for a wide selection of physical and psychological symptoms which occur at the end of the cycle before bleeding occurs and subside as menstruation begins.

Physical symptoms may involve: abdominal bloating, breast tenderness, headache, clumsiness, constipation or diarrhoea, oedema and weight gain.

Emotional/psychological symptoms can include: mood swings, anxiety, depression, panic, irritability, fatigue, insomnia, hostility and anger – sometimes with violence directed at self or others, food cravings and increased appetite and changes in libido. No single therapy has been found to be effective or suitable for all women.

Q3 At the start of the cycle (which is the first day of the menstrual flow) a number of primary follicles begin to develop in the ovary, and initially oestrogen and progesterone levels are low. In the follicular phase, as the follicles develop, oestrogen levels rise considerably and ovulation occurs after 14 days when the follicle ruptures.

Following ovulation, the luteal phase begins: there is an increased production of progesterone and oestrogen as the corpus luteum develops. If the ovum is unfertilized or fails to implant in the endometrium, the corpus luteum degenerates, causing a rapid decline in progesterone and oestrogen to low levels. The cycle then begins again.

Q4 Hypothalamic hormones control the menstrual cycle via negative feedback loops. The hypothalamus produces gonadotrophin releasing hormone (GnRH), which stimulates secretion of the gonadotrophins FSH and LH from the anterior pituitary gland.

FSH stimulates the development of the follicle by activating cell division and secretion of oestradiol. LH rises to a peak and induces ovulation of the maturing ovum. High oestrogen levels cause inhibition of FSH production by a negative feedback mechanism.

In the second half of the cycle oestrogen levels fall and the corpus luteum develops, producing increasing amounts of progesterone and small amounts

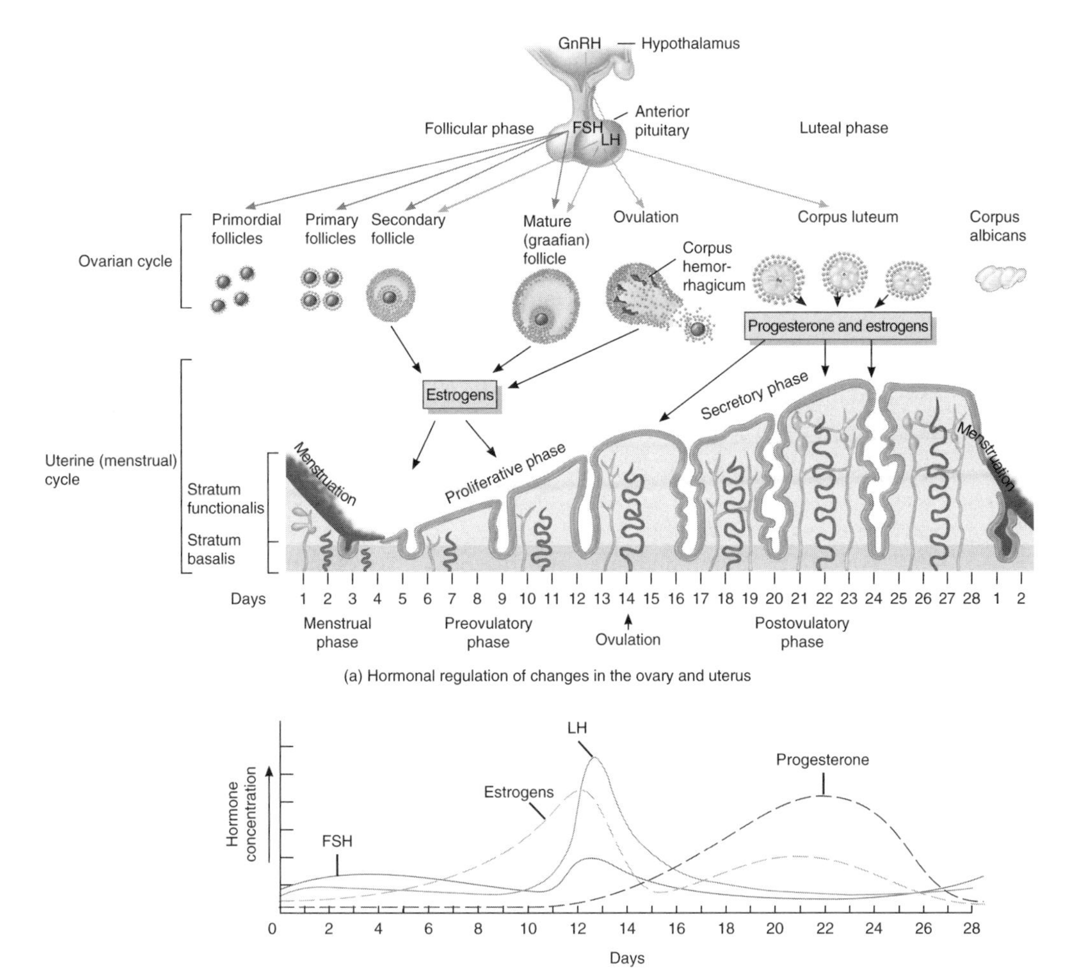

(a) Hormonal regulation of changes in the ovary and uterus

(b) Changes in concentration of anterior pituitary and ovarian hormones

From Tortora and Derrickson *Principles of Anatomy and Physiology, Eleventh Edition 2006*. Reproduced with permission of John Wiley & Sons, Inc.

of oestradiol. The oestrogen and progesterone inhibit further production of LH and FSH by the anterior pituitary gland via negative feedback.

When the corpus luteum degenerates, oestrogen and progesterone levels decrease again, removing the inhibition of LH and FSH production.

Q5 The endometrium contains a mixture of glands and blood vessels embedded in connective tissues; after menstruation has finished it appears thin for a few days. Increasing levels of oestrogen from the follicle then stimulate the endometrium to thicken, the glands hypertrophy and blood vessels become more prominent. This is known as the *proliferative phase*.

After ovulation, progesterone levels rise, the endometrium thickens and becomes more vascular. This is known as the *secretory phase* and normally lasts for 14 days, which allows production of a thick vascular lining for the uterus, suitable for implantation and maintenance of an embryo.

If fertilization and implantation do not occur, oestrogen and progesterone levels fall rapidly and stimulation of the endometrium ceases. Consequently, ischaemic changes occur in the blood vessels and the endometrium is shed, beginning a new cycle.

Q6 Painful or difficult menstruation with severe abdominal cramps is called *dysmenorrhoea*. The symptoms are related to the actions of prostaglandin E_2 and $F_2\alpha$. These prostaglandins are released in the uterus from the phospholipids of the shed cell membranes during menstruation. Prostaglandin E_2 causes vasodilation and platelet degradation. The prostaglandin $F_2\alpha$ causes the contraction of myometrial smooth muscle and is involved in pain sensation. Both oestrogens and vasopressin (antidiuretic hormone (ADH)) can induce the release of prostaglandin $F_2\alpha$. Vasopressin also causes uterine hyperactivity.

In terms of pharmacological management, the level of prostaglandins can be reduced by combined oral contraceptives (progestin-dominant combination oral contraceptives). Analgesics, such as aspirin and paracetamol/acetaminophen, can be used for the pain. Patients should be advised not to take non-steroidal anti-inflammatory drugs (NSAIDs) until the onset of symptoms, as the half-life of released prostaglandins is only a few minutes. Beta-adrenoceptor agonists can reduce the rate and force of uterine contractions. Massage, warming and regular exercise combined with a low-fat diet are also advised.

Q7 Combined oral contraceptives contain synthetic oestrogens and progestogens which inhibit FSH and LH release. These compounds inhibit both ovulation and the secretory phase of endometrial growth. Since they are inhibitors of ovulation, they are able to diminish the symptoms of dysmenorrhoea and premenstrual syndrome, which both appear to be related to the menstrual cycles in which ovulation occurs.

Part 2

Q8 *Endometriosis* is a painful condition in which endometrial cells are found outside the uterus, within the abdominal cavity. Because these ectopic cells are sensitive to the hormonal changes which occur in each menstrual cycle, they proliferate during the cycle in the same manner as the cells lining the uterus. Proliferation of endometrial cells located on organs outside the uterus causes considerable pain and discomfort in the abdomen.

Pregnancy does not eliminate the condition and drug treatment is not always successful.

Q9 Some women's symptoms are helped by non-steroidal anti-inflammatory agents such as ibuprofen, and in some women a combined oral contraceptive

is a satisfactory treatment. The oestrogen component of the combined pill inhibits FSH release and therefore inhibits follicle development. The progesterone component inhibits LH release and, therefore, inhibits ovulation.
An alternative treatment for endometriosis is danazol. Danazol is a modified progestogen which inhibits production of gonadotrophin; it combines androgenic actions with anti-oestrogenic and antiprogestogenic activity.

Q10 When choosing an oral contraceptive, a preparation with the lowest oestrogen and progestogen content that can control the cycle is selected. In older women or women with a history of venous thrombosis who require contraception, a progestogen-only preparation may be more suitable than the combined hormonal type, but unfortunately this is not a suitable choice for Shabana's problem.
One of the risks associated with use of combined hormonal contraceptives is venous thromboembolism, but in women without other contributing risk factors the risk of thrombosis is less than that observed during pregnancy.
The risk of thromboembolism increases with both age and the presence of other risk factors, such as obesity or an immobilizing illness. There is also a small risk of arterial disease, particularly if the woman has a previous history of arterial disease, hypertension, diabetes mellitus or if she is obese.
Shabana is now 39 years old. She has a short stature and at a weight of 13 stones could be considered obese. If she were prescribed a combined oral contraceptive preparation, she would have an increased risk of thromboembolism, which would be further increased if she was also a smoker.

Q11 Menopause marks the end of the reproductive phase of a woman's life which usually occurs between 45 and 55 years of age. Just prior to the menopause, age-related changes occur in the ovary. Only a small number of follicles remain in the ovary and they become less sensitive to FSH and LH. In the years surrounding menopause (the perimenopause) ovulation does not consistently occur during each menstrual cycle and oestrogen levels gradually decrease.
The symptoms which occur around the time of the menopause are associated with reduced oestrogen concentration and can include: hot flushes, sweating, particularly at night, headaches, palpitations, nervousness, anxiety, depression, difficulties in concentration and weight gain.
Skin elasticity decreases and there may be an increase in body hair. In many women there is a significant reduction in bone density, caused by a declining oestrogen concentration, which may lead to osteoporosis and an increased incidence of bone fractures in later life.

Q12 Oestrogens are involved in regulation of menstrual cycle, development of secondary sex characteristics in females and changes in the uterine lining. It affects the metabolism of minerals, carbohydrate, protein and lipid. Oestrogen crosses the cellular membrane and binds to receptors in the nucleus; these receptors are present in both males and females. In females they are located in the hypothalamus, pituitary gland, reproductive system, breast, bone and liver. Oestrogen prevents the reabsorption of bone and increases the build-up of mineral in bones. Oestrogen modifies the metabolism of lipids by reducing the levels of low-density lipoprotein (LDL) and increasing the levels of high-density lipoprotein (HDL) cholesterol. This is thought to reduce the development of coronary artery disease (CAD) and incidence of myocardial infarction (MI).

Q13 HRT is hormone-replacement therapy. Ovarian function decreases at the menopause and the reduction in oestrogen production results in menopausal symptoms such as hot flashes, sweats, vaginal dryness and so on. Natural oestrogens used in HRT decrease these menopausal symptoms. The oestrogens both diminish vasomotor symptoms (e.g. hot flashes, sweating) and inhibit atrophy (e.g. vaginal atrophy), osteoporosis and the incidence of CAD. For women with an intact uterus, the risk of developing endometrial cancer is reduced by including a progestogen in the HRT. The progestogens used in HRT include norethisterone, levonorgestrel and medroxyprogesterone acetate. In addition, HRT is helpful in conditions such as hypogonadism, primary ovarian failure, hypermenorrhoea, endometriosis, acne in adults and also prostate cancer since oestrogen suppresses its main growth factor, that is androgen.

Q14 Oestrogen may be administered orally or by subcutaneous or transdermal routes. In topical administration, first pass metabolism of oestrogens in the liver will be avoided.

Q15 No, women on HRT have the ability to become pregnant. HRT cannot be used as a contraceptive method.

Q16 Some individuals may develop one or more of the following: abdominal cramps, nausea and vomiting, headache, dizziness, depression, mood changes and cardiovascular effects such as hypertension, headache and fluid retention, as well as: uterine fibroids, changes in libido, amenorrhoea, cervical erosion, vaginal candidiasis, breast enlargement and/or tenderness and abnormal secretions and weight gain in long-term use.

Q17 There are some major problems associated with oestrogen-replacement therapy. These include: pulmonary embolism, thromboembolism, seizures, hepatic adenoma and risk of stroke. There is also now evidence of an increased incidence of breast, ovarian and endometrial cancer, which is related to the duration of HRT use. Approximately 14 in every 1000 women aged 50–64 years *not* using HRT develop breast cancer. Use of oestrogen-only HRT for five years in this age group increases the incidence of breast cancer to about 15.5 in every 1000 women; this represents a relatively small increase in risk.

Q18 Oestrogen causes an increase in the growth of endometrial tissue (hyperplasia) in the uterus. This may promote endometrial cancer in long-term use and there are also slightly increased risks of breast and ovarian cancer. However, some women suffer very distressing menopausal symptoms, and HRT has proved to be an effective treatment. The current guidelines for the use of HRT when menopausal symptoms are particularly troublesome recommend that the minimum effective dose of oestrogen is used for the shortest period of time.

Key Points

- In normal 28-day menstrual cycles several primary follicles begin to develop in the ovary at the start of the cycle and by day 14 oestrogen levels have risen and ovulation occurs as one follicle ruptures. Following ovulation, the luteal phase begins, oestrogen and progesterone levels rise as the corpus luteum develops. If fertilization does not occur, the corpus luteum degenerates, causing rapid decrease in oestrogen and progesterone. The cycle then begins again.
- Premenstrual syndrome is associated with many symptoms. Physical symptoms include: abdominal bloating, breast tenderness, headache, clumsiness, constipation or diarrhoea, oedema and weight gain. Emotional and psychological symptoms include: mood swings, anxiety, depression, panic, irritability, fatigue, insomnia, hostility and anger–sometimes with violence, food cravings, increased appetite and changes in libido.
- Dysmenorrhoea is painful or difficult menstruation with severe abdominal cramps. Symptoms are related to the actions of prostaglandins E_2 and $F_2\alpha$, which are released in the uterus during menstruation. Analgesics, for example aspirin, can be used to treat pain; suppression of ovulation using combined oral contraceptives is effective in minimizing these symptoms.
- Endometriosis is a painful condition in which endometrial cells migrate outside the uterus. These ectopic cells are sensitive to hormonal changes and proliferate during the menstrual cycle. Proliferation of endometrial cells outside the uterus causes pain and discomfort in the abdomen. Pregnancy does not eliminate the condition and drug treatment is not always helpful. Ibuprofen may reduce pain and in some women a combined oral contraceptive is effective. Danazol, a modified progestogen with anti-oestrogenic and androgenic actions, may also be satisfactory.
- Menopause, (cessation of menstruation)), usually occurs between 45 and 55 years of age. Prior to menopause, only a few follicles remain in the ovary and they are less sensitive to FSH and LH. Ovulation does not consistently occur and oestrogen levels gradually fall. The symptoms experienced are associated with reduced oestrogen concentration and include: hot flashes, sweating, headaches, palpitations, nervousness, anxiety, depression, poor

concentration and weight gain. Many women suffer from reduction in bone density, leading to osteoporosis and bone fractures in later life.

- HRT usually contains oestrogen plus a progestogen. It is administered either orally or via a skin patch. HRT gives relief from hot flashes and sweating and protects against osteoporosis. The progestogen content protects women with an intact uterus against endometrial cancer. Oestrogen alone may be prescribed for women without a uterus,
- Adverse oestrogen effects can include: abdominal cramps, nausea, vomiting, headache, dizziness, depression, mood changes, hypertension, headache and fluid retention. Some major problems can occur with oestrogen therapy, including thromboembolism, seizures and risk of stroke. There is some evidence of a small increased incidence of breast, ovarian and endometrial cancer, which is related to the duration of HRT use.

CASE STUDY 50 Demi's baby

Part 1

Q1 The primordial follicles in which ova develop, form in the embryo and the maximal number 1–2 million, is present at birth. Follicles are located in the cortex of the ovary and each contains a primary oocyte. Following sexual maturation at puberty about 200 000 follicles remain, some of these will develop into primary follicles. During reproductive life a number of the primary follicles develop further, leading to ovulation, the remainder decline (a process known as *atresia*).

Each month following the menarche, which normally occurs at around the age of 12 years, several follicles begin to develop into Graafian follicles. One of the Graafian follicles becomes dominant, grows very large, produces a bulge at the ovarian surface and finally ovulates. This process is known as the *follicular phase* of the menstrual cycle and is controlled by FSH released from the anterior pituitary gland. FSH release is in turn controlled by the hypothalamus, which secretes a GnRH that stimulates FSH release.

Under FSH stimulation the growing follicle matures and secretes increasing amounts of oestradiol. At about day 14 of the cycle of the average woman, oestradiol triggers a surge of LH secretion from the anterior pituitary gland (a positive feedback effect), which results in the release of the ovum.

The high LH concentration also promotes the formation of a corpus luteum and secretion of progesterone. The function of progesterone from the corpus luteum is to prepare the uterus for possible pregnancy. If pregnancy does not occur, the corpus luteum degenerates about 12 days after ovulation and progesterone levels fall.

In the course of a reproductive life of approximately 40 years, about 500 ovulations will have occurred in the average woman. The end of the reproductive phase of life, menopause, usually occurs around the age of 45–55 years.

Q2 Female fertility can be impaired by abnormalities in ovulation, which are perhaps due to abnormal ovaries or hormone production. There may be anatomical defects in the Fallopian tubes or uterus, production of abnormal cervical mucus or an immunological problem. In young women failure of menstruation and anovulatory cycles can be triggered by a variety of external factors, including a significant weight loss, whether caused by malnutrition, extreme dieting or excessive exercise.

The number of ovarian follicles capable of maturation is fixed at birth. They are depleted by age, and in older women approaching the menopause few follicles remain. At around the time of the menopause (perimenopause) oestrogen levels fall and the ovaries also gradually lose their responsiveness to FSH and LH so that the process of ovulation and fertility is impaired.

Q3 At puberty spermatogenesis begins in the seminiferous tubules: it requires a temperature of 2–3 °C below core body temperature and the low temperature is made possible by the position of the testes in the scrotum. Spermatozoa develop from spermatogonia in the seminiferous tubules. The first mitotic division of these cells produces diploid primary spermatocytes. Some of these cells develop further and undergo meiosis, each producing two secondary spermatocytes. These cells divide again (meiosis) forming haploid spermatozoa, which mature further and are flushed into the epididymis. Following further maturation, the spermatozoa acquire motility and move into the vas deferens. The function of the prostate gland and seminal vesicles is to add fluid to the semen before ejaculation; the ejaculate of the average adult male contains approximately 300 million spermatozoa. During production and maturation of spermatozoa, Sertoli cells facilitate the nutrition of the gametes and the formation of seminal fluid in the seminiferous tubules. At least 60 million spermatozoa are required in the ejaculate for fertility.

Like the control of female reproduction, control of spermatogensis depends on the actions of hormones and negative feedback mechanisms. GnRH from the hypothalamus controls the release of LH and FSH. LH stimulates secretion of testosterone by the testis and FSH stimulates the process of spermatogenesis via actions on Sertoli cells. Testosterone secretion is relatively constant, rather than cyclic, and does not suddenly decrease at a particular stage of a man's life like the oestrogens in females. An increasing testosterone concentration inhibits secretion of GnRH from the hypothalamus and reduces hormonal release from the anterior pituitary. Therefore, spermatogenesis requires adequate secretion of FSH and LH by the pituitary gland, adequate secretion of testosterone, efficient functioning of the Sertoli cells and adequate spermatogonia.

Male fertility is also affected by anatomical problems, which hinder sperm delivery, and by production of inadequate numbers or inadequately functioning spermatozoa. Sperm motility is important for fertility; it is affected by the chemistry of the semen and its viscosity. Approximately 3–7% of men show antisperm antibodies in their semen. If infertility is thought to be due to these antibodies, semen can be collected, washed and then diluted or concentrated according to requirements, to aid conception.

Previous infections, for example mumps, which causes inflammatory changes in the seminiferous tubules, can result in infertility. Males experience age-related changes in hormones and in the reproductive tract; sperm production will decline by approximately one-third by the age of 65 years. However, most elderly men still produce large enough numbers of spermatozoa to remain fertile throughout their life.

Q4 Testosterone determines the characteristics of the male body, promoting the development of male secondary sexual characteristics, the growth of the reproductive tract and development of spermatozoa. Testosterone has multiple anabolic effects. It stimulates the growth of soft tissues and bones, increasing height and weight at puberty and promotes maturation of spermatozoa and continuation of normal spermatogenesis.

Q5 *Pregnancy* refers to the period between fertilization of the ovum by a single sperm to the birth of the baby; the normal gestation period is 40 weeks, starting from the first day of the last menstrual period. Penetration of the plasma membrane of the ovum by a sperm makes it impermeable to further sperms and, following penetration, the genetic material of the two haploid cells fuses to form a diploid zygote. Repeated cell division forms a ball of cells that travels through the Fallopian tube forming a blastocyst which implants in the endometrial layer of the uterus. Implantation results in secretion of chorionic gonadotrophin, which maintains the corpus luteum for the first few weeks of pregnancy; chorionic gonadotrophin peaks at about six weeks. Maintenance of pregnancy requires the secretion of a large concentration of oestrogen and progesterone, and the developing placenta becomes the chief source of these hormones from approximately the third month of pregnancy.

Q6 During pregnancy, there are significant changes in many body systems. Tidal volume and respiratory rate both increase. By the midpoint of pregnancy (at about 20 weeks) blood volume and cardiac output has increased by 30–40% without elevating blood pressure and without a proportional increase in red blood cell production, so anaemia is a fairly common occurrence during pregnancy.

Glomerular filtration rate is increased by about 50% and micturition is more frequent as the enlarged uterus presses on the bladder. Many women become constipated, which is due partly to reduction in gut motility and partly to compression of the large intestine by the growing uterus. Gastric motility is decreased in pregnancy because of the effects of progesterone, and gastric reflux (heartburn) is a common condition.

The nutritional needs of a woman increase during pregnancy, but if a woman does not have an adequate diet the nutrients needed by the foetus are acquired from the mother's body. For example, if the woman's calcium intake is inadequate, calcium will be mobilized from her bones to provide for the developing foetus.

Q7 Following implantation, the cells on the outer surface of the embryo stick to the endometrial surface and the embryo buries itself in the endometrium. The outer layer of membrane surrounding the embryo, the chorion, forms finger-like projections called *chorionic villi*, which produce enzymes to digest the endometrial cells. Capillaries containing foetal blood develop within the villi. Maternal blood carrying oxygen fills the spaces between the villi so that maternal and foetal blood come into close contact. During pregnancy, gases, nutrients and hormones are transferred between foetal and maternal blood. As pregnancy progresses the villi enlarge and the placental structure becomes more complex.

Two umbilical arteries from the foetus carry blood to the placenta and a single umbilical vein returns blood from the placenta back to the foetus. The functions of the placenta in pregnancy are to supply oxygen and nutrients from the maternal circulation to the foetus and to remove waste materials, such as urea and carbon dioxide, from foetal blood.

Part 2

Q8 During pregnancy, contractions of the smooth muscle of the uterus are suppressed because of actions of progesterone. But at the end of pregnancy the concentration of progesterone declines and the high oestrogen concentration increases the density of oxytocin receptors in the myometrium.

Labour begins spontaneously at about 40 weeks and involves actions of oxytocin and prostaglandins. In the early stages of labour the cervix softens and becomes dilated because of the actions of the hormone relaxin, which promotes the dissociation of collagen fibres. Stretching of the cervix triggers a nervous reflex that induces oxytocin secretion from the posterior pituitary gland. At this time the smooth muscle cells of the myometrium have an increased sensitivity to oxytocin, because of the increase in the number of oxytocin receptors at the end of pregnancy. As labour progresses, mechanoreceptors in the cervix provide the sensory element of a positive feedback response, leading to further increase in oxytocin release, so contractions of the myometrium increase in duration, frequency and strength during labour.

Q9 Myometrial contractile activity is modified by both endocrine and autonomic factors. The increase in oestrogens during pregnancy gradually increases both the excitability of uterine smooth muscle and its sensitivity to agonists, particularly oxytocin. The uterus receives sympathetic innervation, which exerts excitatory effects via alpha-1-receptors (α_1-receptors). Uterine smooth muscle also possesses beta-2-receptors (β_2-receptors), which mediate relaxation.

The birth process starts with sporadic contractions of the myometrium. If contractions occur prematurely before 33 weeks' gestation, myometrial relaxants (tocolytic agents) can usually delay the delivery of the baby for 48 hours or more. Beta-2-agonists, such as ritodrine hydrochloride, terbutaline sulfate or salbutamol, are normally given for up to 48 hours to elicit relaxation. Longer treatment is not recommended since there is substantial risk of adverse effects on the mother when treatment is prolonged. A similar reduction in uterine contractile activity can be obtained with calcium channel blocking agents, such as nifedipine.

Atosiban, an oxytocin receptor antagonist, is also licensed for treatment of premature labour. This drug has fewer side effects than the β_2-agonists and is particularly useful if the mother has cardiac disease, because intravenous β_2-agonists can cause adverse cardiac actions in the mother.

Q10 Beta-2-adrenoceptor agonists are used to treat premature labour occurring between 24 and 33 weeks of gestation and allow an extra 48-hour delay in the birth. This delay is usually important for prevention of neonatal respiratory distress; the latter is reduced by administering steroids. Steroids promote the development of surfactant in the foetal/neonatal lung and reduce the incidence of Respiratory Distress Syndrome. However, β_2-agonists can cause maternal side effects in prolonged use, including: tachycardia, palpitations and dysrhythmias, hypotension, chest pain, flushing, vomiting, hypokalaemia and liver problems.

Q11 Labour involves myometrial contractions of increasing strength and regularity. Drugs which stimulate contraction include prostaglandins (e.g. dinoprostone) and oxytocin; these agents can be used both to induce and to enhance labour. Oxytocin is usually administered by intravenous infusion.
Following the delivery of the baby, oxytocin produced from the posterior pituitary gland contracts the uterus, reducing haemorrhage from the uterine arteries.

Q12 Human milk is a bluish-white fluid with approximately 88% water, 6–8% carbohydrate (lactose), 3–5% fat and 1–2% protein. The composition of milk varies from day to day and changes during a single feed: the milk is watery at the start of a feed to satisfy thirst, but the fat content of the milk increases towards the end of the feeding period.
Several anti-infective substances are present in breast milk: lysozyme, complement, and immunoglobulins, particularly IgA. IgA is present at even higher concentration in the first secretion from the mammary gland, colostrum. IgA is thought to adhere to the infant intestinal mucosa to form a kind of immunological barrier to infectious agents, such as rotaviruses. The anti-infective substances are very important to infant health as the baby's immune system is immature at birth and it is unable to produce antibodies for several months. Living cells, lymphocytes and phagocytic cells, such as macrophages, are also present in human milk.
A further important constituent of human milk, which is not present in formula milk, is lactoferrin, which complexes iron in milk so reducing the iron available for gram-negative bacteria to multiply in the infant gut.
There are many benefits of breast milk to the baby. Breastfed infants have a lower incidence of infections of the ear, gut, respiratory and urinary systems than bottle-fed infants. There is automatic compensation for fluid losses: in hot weather the baby will take more milk but the milk becomes more dilute, so baby does not consume extra calories. Obviously, a major advantage of breast milk is that it is free and is conveniently available on demand, without the need for equipment to sterilize bottles.

Q13 A mature mammary gland consists of lobules containing clusters of alveoli that produce the milk. A duct drains each cluster of alveoli and several ducts fuse to form lactiferous or mammary ducts which take milk to the nipple. The breasts grow during pregnancy but milk production is inhibited until birth is completed because of the high concentration of oestrogen and progesterone present during pregnancy. When the level of these hormones declines at birth, prolactin from the anterior pituitary gland promotes milk production. Secretion of the milk is initiated by the infant's suckling, which stimulates oxytocin release from the posterior pituitary. Oxytocin promotes contraction of the smooth muscle in the alveoli, which makes the milk available to the infant. Suckling also promotes a surge in prolactin secretion, which initiates further milk production by the alveoli, ready for the next feed.

Q14 The secretion of prolactin from the anterior pituitary gland is stimulated by suckling and this hormone inhibits GnRH secretion from the hypothalamus. A rise in GnRH normally causes release of FSH and LH from the anterior pituitary. When GnRH is inhibited, the secretion of FSH and LH is reduced and as a result secretion of oestrogen and progesterone falls and follicular development and ovulation is suppressed. This protection does not continue indefinitely: usually, breast feeding inhibits ovulation for six months or maybe more, but eventually ovulation occurs again. Lactation is a useful, but not a reliable, method of birth control.

Key Points

- A large number of ovarian follicles exist at birth and each contains a primary oocyte. Following sexual maturation, some develop into primary follicles, leading to ovulation; the remainder decline. Each month several follicles begin developing into Graafian follicles. One of these dominates, grows very large and will finally ovulate. The follicular phase is regulated by FSH, a process controlled by the hypothalamus, which secretes GnRH. At day 14, oestradiol triggers a surge of LH secretion from the anterior pituitary gland and an ovum is released. High LH concentration promotes formation of a corpus luteum and secretion of progesterone to prepare the uterus for pregnancy. If this does not occur, the corpus luteum degenerates and progesterone levels fall.
- At puberty, spermatogenesis begins in the seminiferous tubules. It requires a temperature of 2–3 °C below core body temperature. Spermatozoa develop from spermatogonia in the seminiferous tubules. Mitotic division produces diploid primary spermatocytes, some of which develop further and undergo meiosis, each producing two secondary spermatocytes. These cells divide again, forming haploid spermatozoa, which enter the epididymis. Following further maturation, spermatozoa become motile and move to

the vas deferens. The prostate gland and seminal vesicles add fluid to semen before ejaculation.

- Male infertility: at least 60 million spermatozoa are required in an ejaculate for fertility. Male fertility is affected by anatomical problems, which can hinder sperm delivery, and by production of inadequate numbers or inadequately motile spermatozoa. Approximately 3–7% of men show antisperm antibodies in their semen. Previous infections, for example mumps, which causes inflammation in the seminiferous tubules, can result in infertility.
- Female infertility: this can be due to abnormalities in ovulation or abnormal hormone production or ovaries. There may be anatomical defects in the Fallopian tubes or uterus, production of abnormal cervical mucus or an immunological problem. In young women failure of menstruation and anovulatory cycles can be triggered by many external factors, including rapid weight loss. Since the number of ovarian follicles capable of maturation is fixed at birth, few remain as women approach the menopause. The oestrogen levels also fall and ovaries lose their responsiveness to FSH and LH, and so ovulation and fertility is impaired in older women.
- Fertilization of an ovum prevents further sperm penetration. Repeated cell division forms a ball of cells that travels through the Fallopian tube and implants in the endometrium. Implantation elicits secretion of chorionic gonadotrophin, which peaks at about six weeks and maintains the corpus luteum for the first weeks of pregnancy. Maintenance of pregnancy then requires increased secretion of oestrogen and progesterone, and the developing placenta is the chief source of these hormones from the third month of pregnancy.
- Labour begins spontaneously at about 40 weeks and involves myometrial contractions of increasing strength and regularity, which are due to actions of oxytocin and prostaglandins. Drugs which can stimulate contraction include prostaglandins (e.g. dinoprostone) and oxytocin; these agents can be used both to induce and enhance labour. Oxytocin is usually administered intravenously.
- If labour begins prematurely, before 33 weeks' gestation, myometrial relaxants (tocolytic agents) can delay it for 48 hours. Beta-2-agonists, such as ritodrine hydrochloride, are given for up to 48 hours to elicit relaxation. Longer treatment is not recommended, because of risk of adverse effects on the mother. Nifedipine and the oxytocin receptor antagonist atosiban are also used to reduce premature uterine contractile activity.
- Breasts grow during pregnancy but milk production is inhibited until after birth by the high concentration of oestrogen and progesterone present

during pregnancy. When these hormones decline at birth, prolactin from the anterior pituitary gland can promote milk production. Secretion of milk is initiated by the infant's suckling, which stimulates oxytocin release from the posterior pituitary and contraction of smooth muscle in the alveoli. Suckling also promotes the surge in prolactin secretion, which initiates further milk production ready for the next feed.

- Human milk contains water, lactose, fat and protein. The composition of milk varies from day to day and changes during a single feed: milk is watery at the start of a feed, but the fat content increases towards the end. Several anti-infective substances are present in colostrum and breast milk: lysozyme, complement, immunoglobulins, particularly IgA, and lactoferrin, which complexes iron in milk to reduce the iron available for gram-negative bacteria to multiply in the infant gut. Living cells, lymphocytes and phagocytic cells, such as macrophages, are also present in human milk.

Glossary

ACE inhibitor angiotensin-converting enzyme inhibitor used for hypertension and congestive heart failure.

ADHD attention-deficit hyperactivity disorder, which can be a chronic condition associated with hyperactivity, forgetfulness, distractibility and poor control. This condition is mostly seen in children but can continue into adulthood.

ADD attention-deficit disorder. Patients with this disorder have problems in focusing.

Addison's disease an endocrine disorder associated with the adrenal gland, when low levels of steroid hormones are produced.

Adrenaline a hormone/neurotransmitter that is released under certain conditions, e.g. in 'fight or flight'.

Adrenal cortex the outer layer of the adrenal gland which is involved in the production of mineralocorticoids and glucocorticoids, eg aldosterone and cortisol. It is also involved in the production of androgen.

Adrenal medulla the inner layer of the adrenal gland which is involved in the production of dopamine, adrenaline and noradrenaline.

Adrenocortical hormones hormones secreted by adrenal cortex.

Albumin a blood plasma protein which is produced by the liver and is involved in maintaining osmotic pressure.

Aldosterone a hormone produced by the adrenal cortex which is involved in the regulation of sodium and potassium in the blood.

Clinical Physiology and Pharmacology Farideh Javid and Janice McCurrie

Alkalosis a condition which is associated with a reduction in the level of hydrogen ions in the arterial blood. This causes the pH of the blood to increase.

Allegra an antihistamine drug that is used for allergic conditions, e.g. hayfever.

Allergic rhinitis an allergic condition associated with sneezing, runny nose and itchy eyes.

Alpha-blocker a drug that blocks alpha adrenoceptors.

Alpha (α) **gliadin** found in wheat and other cereals, a peptide component of gluten.

Alzheimer's disease a neurodegenerative disease which is a type of dementia. It is associated with a progressive decline in cognition.

Amantadine a drug used for Parkinson's disease. It also has antiviral properties.

Amitriptyline hydrochloride a tricyclic anti-depressant drug which inhibits serotonin and noradrenaline re-uptake.

Anaemia a deficiency in the red blood cells or haemoglobin.

Aneurysm a bulge or dilated blood vessel.

Angina a condition associated with a sharp pain which is due to the contraction or tightness of an area in the body.

Angina pectoris a condition associated with a sharp pain in the chest due to insufficient/lack of blood in the heart.

Angiotensin an oligopeptide which is produced from angiotensinogen (which in turn is made in the liver). It can cause the blood vessels to contract, leading to an increase in blood pressure. It also initiates the release of aldosterone from the adrenal cortex.

Antacid any drug or chemical that neutralizes stomach acidity.

Anticholinesterase any drug that can prevent the action of the enzyme responsible for converting acetylcholine to choline and acetic acid.

Antidiuretic hormone (ADH) a hormone which is responsible for the regulation of water retention in the body. It is also called arginine vasopressin (AVP) or argipressin.

Antihypertensives drugs used for high blood pressure.

Antiplatelet agents drugs that are used to prevent thrombus formation by reducing platelet aggregation.

Anxiolytics drugs that are used in patients with anxiety, e.g. benzodiazepines.

Amylase an enzyme responsible for breaking down starch into glucose.

Aorta the largest artery which is connected to the heart from the left ventricle. It carries oxygenated blood from the heart to the other parts of the body through the systemic circulation.

Atrial fibrillation abnormal rhythm of the atria in the heart.

Aspirin a drug that has been used to reduce pain, fever and inflammation. If used for a long time at low doses, it has antiplatelet activity in blood and prevents coagulation.

Asthma a respiratory condition associated with constriction of the airways.

Atelectasis a condition when the lungs are collapsed with no air in the alveoli.

Atheroma a fatty material deposited inside an artery.

Atherosclerosis an inflammatory chronic disease associated with the walls of the arterial blood vessels.

Atrial natriuretic peptides (ANP) a hormone that is released by cardiac cells following a high blood pressure. It is involved in the control of water, sodium and adiposity. It is also known as atrial natriuretic factor (ANF) and atriopeptin.

Atropine a muscarinic antagonist.

Azelastine hydrochloride an antihistamine which is used for allergic reactions such as hay fever and allergic conjunctivitis.

Beta-2-agonist (β_2 agonist) a drug that binds to β-2-adrenoreceptors.

Beta-blocker (β-blocker) a drug that blocks β-adrenoreceptors.

Benzodiazepine psychoactive drugs that are used for anxiety, insomnia, muscle spasms and seizures.

Bilirubin a yellow/orange pigment derived from the breakdown of haemoglobin, that is excreted in bile.

Biguanide a drug used in diabetes that reduces the level of glucose in the blood.

Bronchoconstrictor a drug that causes constriction of the airways.

Bronchodilator a drug that causes dilation of the airways.

Captopril a drug that is used for hypertension and congestive heart failure. It is an angiotensin-converting enzyme (ACE) inhibitor.

Carbamazepine an antiepileptic drug.

Carbidopa a drug used in Parkinson's disease which prevents the metabolism of levOdopa (L-dopa) in the periphery.

Carbimazole a pro-drug that is converted to its active form, methimazole, used in hyperthyroidism.

Cardioversion a direct current (DC) shock across the chest. This depolarises the heart, allowing an organised rhythm to develop.

Cardiac arrest failure of the heart to contract owing to inadequate, or the stopping of, circulation of blood during systole.

Carotid arteries the arteries that carry the oxygenated blood to the head and neck.

Celiac disease (coeliac disease) a condition associated with diarrhoea, pain, weight loss (particularly in children) and fatigue due to sensitivity of the intestinal mucosa to the protein gluten. It is classed as an autoimmune disease of the small intestine.

Chlortalidone a diuretic (thiazide) drug used in hypertension.

Chronic bronchitis inflammation of the bronchi for a long period e.g. two years, which could be part of chronic obstructive pulmonary disease (COPD).

Colostrum the first secretion from the mammary gland following birth. Colostrum is rich in protein and antibodies which confer passive immunity on the newborn.

Congestive heart failure a condition, associated with problems of filling or pumping in which the heart cannot maintain an adequate circulation of blood to meet the metabolic demands of the tissues.

COPD or chronic obstructive pulmonary disease, a term used for a group of conditions/diseases associated with obstruction to expiratory flow, eg bronchitis, emphysema.

Coronary circulation the blood supply to heart muscle.

Cortisol or stress hormone, is released by the adrenal gland (the cortex); it increases the level of glucose in the blood and also blood pressure. It also suppresses the immune system in the body.

Creatinine is the byproduct of creatine phosphate. The reaction takes place in the muscle.

Creatine phosphokinase an enzyme responsible for converting creatine to phosphocreatine in muscle. This enzyme increases and peaks within 24 hours of myocardial infarction.

Cromoglicate a drug that is used for allergic conditions, such as allergic rhinitis, asthma, conjunctivitis etc. It prevents the release of histamine from mast cells.

Cushing's disease is an endocrine disorder caused by an increase in the level of cortisol in the blood.

Cyanosis a condition due to sluggish blood flow and accumulation of deoxygenated haemoglobin in the tissues. Cyanosis gives tissues a blue colouration, best seen in the lips and mucous membranes.

Cystic fibrosis a condition that affects the lungs, pancreas and the digestive system. Patients develop difficulty in breathing, experience frequent infection, malabsorption and diarrhoea; growth will be affected.

Deep-vein thrombosis (DVT) refers to a condition when there is a clot in a deep vein. This usually occurs in veins of the thigh.

Dementia refers to a disease associated with damage or dysfunction of different brain areas, which causes progressive loss of cognitive function.

Desferrioxamine an iron chelating agent used in iron poisoning.

Desmopressin a drug that acts similarly to antidiuretic hormone.

Detrusor is the smooth muscle of the urinary bladder.

Dexamethasone a drug that mimics the action of glucocorticoid and is used for inflammation and autoimmune conditions.

Diabetes mellitus a condition associated with high levels of glucose in the blood caused by inadequate amounts of insulin or ineffectiveness of insulin in the body.

Diabetes insipidus a condition that is associated with failure of the kidneys to concentrate urine due to reduction in the level of antidiuretic hormone. Patients lose considerable amounts of dilute urine even when the fluid intake is decreased.

Diamorphine an opioid analgesic agent that is synthesized from morphine.

Digoxin a cardiac glycoside used for different heart conditions, such as fibrillation.

Diverticula pouches which form in the wall of the colon and which may become infected and inflamed.

Diverticulitis a digestive condition associated with the formation of pouches on the colon.

Donepezil or Aricept, is used in diseases such as Alzheimer's and inhibits the action of acetylcholinesterase enzyme activity.

Dopamine a hormone/neurotransmitter released mainly from the substantia nigra in the brain. Low levels of dopamine are associated with the development of Parkinson's and high levels are associated with the development of schizophrenia.

Dysmenorrhoea a condition associated with painful menstruation. The pain is concentrated in the lower abdomen or pelvis.

Dyspnoea a condition associated with difficulties with breathing. This may be described as 'shortness of breath' or pain when breathing.

Economy-class syndrome tendency of blood to clot following prolonged inactivity on long haul flights, which leads to deep vein thrombosis.

ECG or electrocardiogram, is the trace of the heart's electrical activity. The record is obtained from 12 sites at various agreed positions on the chest and limbs.

Endometriosis a condition associated with the abnormal growth of the lining of the uterus. Endometrial tissue migrates from the uterus to other abdominal structures and elicits pain when its growth is stimulated during the menstrual cycle.

Embolism a condition when blood vessels are blocked, e.g. by a clot or a bolus of fat carried to the site by blood flow.

Epigastric pain abdominal pain which in severe conditions may be a sign of heart problems such as infarction.

Epilepsy a neurological disorder that is associated with seizures.

Epinephrine or adrenaline, is a 'fight or flight' hormone released by the adrenal gland. It is also classified as a neurotransmitter.

Erythrocyte sedimentation rate is a blood test in which the extent to which red cells in a blood sample settle to the bottom of a tube is timed. A changed rate can indicate inflammation.

Erythropoiesis production of red blood cells (erythrocytes).

Erythropoietin a hormone produced in the kidney which stimulates red bone marrow to produce erythrocytes.

Essential hypertension the most common type of high blood pressure (hypertension) in which the cause of the high pressure is unknown.

Exophthalmos protrusion of the eyeballs with limitation of eye movement.

Exudate Fluid derived from a site of injury.

Ferritin a complex which consists of 24 protein sub-units: iron is stored as ferritin.

Ferrous gluconate an absorbable iron salt, used as an iron supplement.

Fexofenadine a drug that is classed as an antihistamine and is used for allergic reactions, such as hay fever.

FEV_1 is the forced expiratory volume in one second.

Flatus gas in the intestinal tract.

Fluoxetine a selective serotonin reuptake inhibitor (Prozac) used to treat depressive illness.

Fluphenazine a neuroleptic drug.

FSH follicle stimulating hormone. A hormone released from the anterior pituitary gland which stimulates ovarian follicles in the female and permatogenesis in males.

Furosemide a type of diuretic drug which greatly increases salt and water excretion by acting on the loop of Henle in the nephron. It is used for conditions such as oedema and heart failure.

FVC or forced vital capacity, which is the maximum amount of air that can be expired after a maximum inhalation.

GABA gamma amino butyric acid, an inhibitory neurotransmitter.

Glaucoma a disease associated with damage to the optic nerve. It can be caused by an increase in intraocular pressure and can cause blindness if not treated.

Glibenclamide an oral hypoglycaemic agent used in treatment of diabetes mellitus.

Glomerulus a network of capillaries in nephrons that is surrounded by Bowman's capsule.

Glomerulonephritis a renal disease that is associated with inflammation of the glomeruli.

Glossitis inflammation of the tongue.

Glucagon a hormone that is released from the pancreas and is involved in the production of glucose from glycogen in a hypoglycaemic condition.

Glucocorticoids steroid hormones involved in the metabolism of glucose.

Gluconeogenesis production of glucose from non-carbohydrate source.

Glucose tolerance test, a test indicating the extent of glucose utilisation. This is impaired when insulin secretion is inadequate.

Gluten-sensitive enteropathy or celiac disease, is an autoimmune disease of the small intestine. The symptoms are chronic diarrhoea and fatigue; children fail to thrive.

Glyceryl trinitrate a prodrug that is used for heart conditions such as angina and heart failure.

Glycogenolysis the process of conversion of glycogen to glucose-6-phosphate, which is initiated by glucagon and epinephrine.

Grave's disease is a disorder of the thyroid gland and is associated with hyper-thyroidism, or goiter.

Haloperidol an antipsychotic drug that is used in schizophrenia.

Haematocrit the amount of red blood cells in the blood.

Haematuria a disease of the kidneys and/or urinary tract, which is associated with pink urine caused by the existence of red blood cells in the urine.

Haemoglobin a protein that contains iron and oxygen in the blood.

Haemorrhage loss of blood.

Haemoptysis production of blood-stained sputum.

HCl hydrochloric acid.

Heartburn a burning sensation caused by the leakage of acid from the stomach to the lower part of the oesophagus. The name has no connection to the heart.

Heart attack or myocardial infarction, is associated with damage to the heart due to blockade or insufficient blood flow to the heart.

Heart failure a condition when the heart cannot be filled and/or cannot pump out enough blood to the body.

Heart rate is the number of beats per minute, which is in the range of 100–150 for infants, neonates and children and much lower in adults, 70 and 75 beats per minute in males and females respectively.

Heart sounds are due to movements of blood and closure of the heart valves during each cardiac cycle. The two prominent sounds are 'lub', due to closure of the A-V valves, and 'dup' due to closure of the semi-lunar valves in the aorta and pulmonary artery.

Heparin an anticoagulant agent. It is poorly absorbed orally and must be given subcutaneously or intravenously.

Histamine a neurotransmitter that is released from the mast cells, the brain and enterochromaffin cells in the stomach. It is released during an injury and inflammation.

Histamine H-2 antagonists (H_2 antagonists) drugs such as cimetidine that block type 2 histamine receptors, which in turn reduce/block acid secretion in the stomach.

Hormone-replacement therapy a treatment to overcome the reduction of the release of oestrogen and progesterone in menopausal (pre- and post-menopausal) women.

Hypercalcaemia increase in the amount of calcium in the blood from 9–10.5 to 12–16 mg dl^{-1}.

Hyperkalaemia increase in the level of potassium in the blood (higher than 5.0 mmol l^{-1}).

Hyperglycaemia increase in the level of glucose in the blood (higher than 120 mg dl^{-1} in a fasting adult).

Hypertension a condition associated with a high blood pressure of 140–90 mmHg (systolic and diastolic) or above. The normal blood pressure is 120–80 mmHg for systolic and diastolic blood pressure.

Hyperthyroidism a condition when the thyroid gland will become overactive, which will result in higher levels of thyroxine (T4) and triiodothyronine (T3).

Hyperparathyroidism a condition when the parathyroid glands will become overactive, which will result in higher levels of parathyroid hormone, which in turn leads to an increase in the level of calcium and a reduction in the level of phosphorus in the blood.

Hypocalcaemia a condition when the level of calcium is reduced in the body to levels lower than 2.2 mmol l^{-1}.

Hypoglycaemia a reduction in blood glucose below normal. Acute hypoglycaemia can cause unconsciousness and is a medical emergency.

Hypomagnesia a condition when the level of magnesium is reduced in the body to levels lower than 0.7 mmol l^{-1}.

Hypotension reduction in blood pressure.

Hypothyroidism a condition when the thyroid gland will become less active, which will result in low levels of thyroxine (T4) and triiodothyronine (T3).

Hypothalamus a brain area located between the thalamus and the brain stem. It is involved in regulating, processing and controlling many different functions and activities, such as secretion of hormones, autonomic functions, hunger, temperature etc.

Hypovolaemia a condition when there is a reduction in the amount of blood volume.

Hypoxia a condition associated with a reduction in the level of oxygen in the blood.

Iatrogenic a condition or illness caused by a drug or a medical treatment or procedure.

Infarct or infarction, occurs when parts of tissues in an organ do not receive enough blood, this leads to tissue damage or death.

Inferior vena cava the vein that is responsible for carrying low oxygenated blood from the body into the heart.

Inflammation the response of the body to injury or infections, characterised by heat, swelling, redness and pain.

Inflammatory bowel disease a collective term for a group of diseases associated with the inflammation of the intestine.

Ileum is the terminal part of the intestine (2–4 m in length) following the jejunum.

Insulin a hormone responsible for many functions in the body, such as transforming glucose and fat into glycogen and triglycerides. It is secreted by the special cells of islets of Langerhans located in the pancreas.

Insulin-dependent diabetes a condition associated with insufficient amounts of insulin or failure of insulin to take up glucose from the blood.

Intestinal flora useful microorganisms which are present in the gastrointestinal tract. They have many functions, including producing hormones and vitamins.

Ischaemia damage of tissue as a result of an insufficient supply of blood.

Isocarboxazid an irreversible inhibitor of the enzyme, monoamine oxidase.

IV intravenous.

Jaundice a condition when a high level of bilirubin (higher than 2 mg dl^{-1}) in the blood causes the skin, the whites of the eyes and the mucous membrane to become yellow.

Jejunum the middle section of the small intestine (2.5 m in length) following the duodenum.

Ketoacidosis a condition where the pH of the blood decreases following the formation of the ketone bodies, which in turn are made by the breakdown of fatty acids.

Labetolol a drug that is classified as an adrenergic antagonist, which blocks alpha-1 and beta-receptors and is used to treat hypertension.

Lactulose a semi-synthetic disaccharide which is not absorbed in the intestine. It is used to treat constipation.

Laxative any substance that can increase the motility of the intestine by increasing bulk, stimulating the intestinal muscles or by other mechanisms. They are used for constipation.

L-dopa or levodopa, is used in Parkinson's disease to increase the level of dopamine in the body.

LH or luteinizing hormone, is produced by the anterior pituitary gland. It stimulates ovulation, secretion of progesterone and prepares the breasts for milk secretion.

Libido Sexual interest or desire.

Lipase enzyme responsible for the metabolism of lipid in the body.

Lithium first-line medication for manic depression.

Loop diuretics drugs that can reduce blood pressure and oedema by acting on the loops of Henle in the kidney. They are classified as diuretics.

Loperamide hydrochloride a drug used for diarrhoea. It is classified as an opioid agonist.

Losartan an angiotensin 2 receptor antagonist.

Malathion an organophosphate that prevents the action of cholinesterase enzyme.

Mania an extreme condition when the mood and energy are elevated and patients experience strange thoughts. Patients may experience alternate episodes of depression.

Manic depressive disorder a disorder associated with alternate episodes of mania and depression. It is classed as bipolar disorder.

Mean corpuscular volume a useful value of average red blood cells to classify patients with anaemia. It is the product of haematocrit divided by the number of red blood cells.

Menopause the cessation of the menstrual cycle.

Menarche age / time of the first menstrual period.

Menorrhagia a condition when the menstruation is very heavy and its duration is long.

Menstrual disorders any disorder associated with abnormal menstruation.

Methylphenidate a drug used for the treatment of attention deficit disorder or attention deficit hyperactivity disorder.

Methylxanthines (aminophylline and theophylline) are smooth muscle relaxants used in asthma treatment.

Migraine a neurological disorder that is associated with recurrent, severe headache. Some patients may also experience visual disturbances and nausea.

Miosis a condition associated with the constriction of the pupils.

Misoprostol a prostaglandin analogue which promotes healing of gastric ulcers.

Mitral valve a valve between the atrium and the ventricle in the left side of the heart. It is also called left atrioventricular valve or bicuspid valve.

Mitral stenosis a heart condition in which the mitral valve is narrowed and incompetent.

Monoamine oxidase inhibitors a group of drugs which decrease the activity of the enzyme, monoamine oxidase, and cause the accumulation of amine neurotransmitters.

Montelukast a leukotriene receptor antagonist used in asthma prophylaxis.

Mucolytics drugs which reduce the viscosity of mucus in the respiratory tract.

Myasthenia gravis a neuromuscular disorder associated with extreme muscle weakness and fatigue. It is classified as an autoimmune disorder.

Mydriasis a condition associated with the dilation of the pupils of the eyes. It is opposite to miosis.

Myocardial infarction or heart attack, is associated with a reduction/block of blood supply to the tissues of the heart, which leads to ischaemia and damage.

Myometrium the middle layer of the wall of the uterus. It is composed of smooth muscle, vascular cells and supporting tissues.

Myxoedema a condition associated with thyroid hormone deficiency in which mucopolysaccharides accumulate in subcutaneous tissues.

Nebuliser a device to convert drug solutions or suspensions into an aerosol for inhalation. This allows relatively high concentrations of drug to reach small airways.

Neostigmine a drug that acts as an inhibitor of cholinesterase enzyme, which in turn is involved in breaking down acetylcholine.

Nephron unit of the kidney responsible for filtering the blood, reabsorbing and excreting materials/ions.

Neuroleptics (antipsychotics) tranquilisers which quieten disturbed patients without impairing consciousness.

Neurotransmitters chemicals stored in vesicles in nerve axons which are released in response to action potentials.

Neutropenia a reduction in circulating polymorphonuclear leakocytes.

Nifedipine a dihydropyridine calcium channel blocking drug.

NMDA (N-methyl-D-aspartate) a type of glutamate receptor.

Noradrenaline or norepinephrine, a neurotransmitter released from the sympathetic nervous system and also in the central nervous system; also a hormone that is released from the adrenal gland. It is also involved in 'fight or flight' response.

NSAIDs non-steroidal anti inflammatory drugs, eg aspirin.

Obsessive–compulsive disorder a disorder associated with obsessive behaviour or thoughts or repetitive compulsive actions such as washing or checking.

Obstructive lung condition any disease associated with the obstruction of the airways in the lung which limits expiratory airflow.

Oedema a condition associated with the accumulation of fluid in tissue.

Oesophagus a tube that connects the pharynx to the stomach; food is passed through this muscular tube to the stomach.

Oestrogen the major type of female steroid hormone.

Oral contraceptives pills that prevent pregnancy. There are two different types: the combined pill which consists of both oestrogen and progesterone, and the progestogen-only pill which consists of only progestogen.

Orthostatic hypotension or postural hypotension, which is a reduction in blood pressure following a change in the posture from lying to standing up.

Palpitation an increased awareness of the heartbeat, particularly when the rhythm becomes fast or irregular.

Pancreatitis a condition when the pancreas becomes inflamed.

Paraesthesia an abnormal skin sensation such as burning, tingling, numbness with no apparent cause.

Paranoia a condition associated with thought disturbance.

Parasympathetic one of the divisions of the autonomic nervous system.

Parathyroid glands glands which are situated behind the thyroid glands in the neck. They produce parathyroid hormone.

Parkinson's disease a neurodegenerative disorder associated with bradykinesia and akinesia; speech and motor skills will be affected. This condition is due to a deficit in the dopaminergic pathway.

Pepsinogen a digestive enzyme released from the stomach. It is responsible for the metabolism of proteins.

Peptic ulcer ulcer of the digestive system caused by actions of the acid gastric secretions.

Phenelzine irreversible inhibitor of the enzyme monoamine oxidase.

Phenylephrine a vasoconstrictor drug used to reduce vascular congestion. It is classified as an agonist of alpha 1-adrenoreceptors. It also causes dilatation of the pupils and increases blood pressure.

Phosphodiesterases a family of enzymes which inactivate cyclic nucleotides, cAMP and cGMP, to 5'AMP and 5'GMP respectively by hydolysis.

Pilocarpine a drug used for glaucoma. It is classified as a muscarinic receptor agonist.

Pituitary gland a gland located at the base of the brain under the hypothalamus and is stimulated by the hormones released from the hypothalamus. It is divided into anterior and posterior sections.

Phenylephrine a drug used to reduce congestion. It is classified as an agonist of alpha-adrenoreceptors. It also causes dilation of the pupils and enhances blood pressure.

Plasminogen plasmin is released to the blood as plasminogen and is involved in breaking down blood clots.

Platelet or thrombocytes, are involved in blood clotting.

Pneumothorax presence of air in the chest, e.g. following a stab wound, which causes the chest to collapse.

Polydipsia a condition when patients experience constant thirst in spite of drinking loads of fluids. This condition happens in diabetes.

Polyphagia increased hunger.

Polyuria a condition when patients need to urinate excessively. This condition happens in diabetes.

Postural drainage draining the respiratory tract of secretions following the positioning of the body below the area that needs draining.

Pralidoxime is a substance used to inactivate poisoning by organophosphates or acetylcholinesterase inhibitors.

Prazosin a receptor antagonist drug that binds to alpha-adrenergic receptors and is used for conditions such as hypertension.

Pre-eclampsia abnormal increase in blood pressure during pregnancy.

Premenstrual syndrome a group of conditions/disorders associated with the menstrual cycle.

Progesterone a steroid female hormone.

Propranolol a receptor antagonist drug that binds to beta-adrenoceptors. It is used for hypertension.

Proteinuria a condition associated with the presence of protein in the urine.

Proton pump inhibitors a group of drugs that are very effective in reducing gastric acid secretion.

Ptosis drooping of the upper eyelid.

Pulmonary embolism occlusion of pulmonary artery by a clot.

Pylorus an area of the stomach just before the duodenum.

Ranitidine a drug that is used in patients with peptic ulcer. It inhibits the secretion of acid in the stomach. It blocks the histamine H-2 receptors.

Renal artery stenosis a condition when the renal artery is narrowed.

Renin a proteolytic enzyme which converts angiotensinogen into angiotensin 1.

Renin–angiotensin is a system which is involved in the long-term control of blood pressure.

Residual volume (RV) the amount of air in the lungs that cannot be expired, even after a maximal and forceful expiration.

Reticuloendothelial system a system of different cells such as macrophages, monocytes, cells of reticular connective tissues that are involved in phagocytosis, histiocytes and some cells from the liver (called Kupffer cells).

Rheumatoid arthritis an autoimmune disease that makes joints inflamed and painful.

Rheumatic fever an inflammatory condition which may occur following strep throat or scarlet fever. Different organs may also be affected, such as the joints, heart, skin and the brain.

Rhinorrhoea or runny nose, occurs following a cold or any allergic reaction or hay fever.

Risperidone atypical neuroleptic agent used to treat schizophrenia and psychotic conditions.

Salbutamol a drug that is classed as beta-2-receptor agonist and is used for asthmatic patients to relax the airways.

Selective serotonin re-uptake inhibitors a group of drugs used mainly as antidepressants which increase the level of serotonin in the synapse. They may also be used for anxiety.

Sex steroid sex hormones responsible for the development of primary and secondary sexual characteristics.

Schizophrenia a mental disorder associated with change of personality, perception and severe paranoia.

Seizure or 'fit', a condition associated with the propagation of abnormal electrical activity of brain cells. Patients go into convulsion and abnormal movement, and some have symptoms of psychiatric disorders.

Sodium nitroprusside a drug that causes dilation of blood vessels as a result of producing nitric oxide.

Spermatogenesis the process of production of sperm in males.

SSRI a selective serotonin (5HT) re-uptake inhibitor.

Steatorrhoea formation of floaty, bulky and pale stool as a result of malabsorption in some digestive diseases.

Sternum a flat bone situated between the right and left ribs at the centre of the thoracic cage.

Stroke a condition when a blockade, or insufficient supply, of blood to the brain cells causes a failure in brain function.

Sulfonylurea drugs that are used in diabetes which increase the secretion of insulin from the pancreas.

Sumatriptan a 5-HT-1D receptor agonist which is used for migraine.

Sympathetic anything associated with the sympathetic nervous system, which is one of the divisions of the autonomic nervous system.

Synapse a tiny gap between a neurone and another cell where impulses pass from cell to cell via release of a neurotransmitter.

Tachycardia an increase in the heart rate.

Tachypnoea an increase in the breathing rate.

Terbutaline a short-acting b2 agonist used in treatment of asthma.

Therapeutic index the ratio between the minimum effective dose and the maximum tolerated dose (although the latter may be a toxic dose to some individuals).

Testosterone a steroid hormone produced by the testes and ovaries in males and females respectively.

Thiazide a diuretic drug which inhibits the reabsorption of sodium and chloride ions from the tubules in the kidney.

Thiazolidinedione a drug which decreases insulin resistance in tissues and so reduces blood glucose concentration.

Thrombolysis a process of destroying the clot formed in the blood.

Thrombolytic agents drugs used to destroy clots in the blood.

Thrombosis Formation or presence of a clot in the vascular system or heart.

Troponin a combination of three proteins involved in the contraction process of skeletal and cardiac muscles.

Thyroid gland located in the neck and involved in the processes of metabolism, growth and functions of different organs. It produces hormones such as triiodothyronine (T3), thyroxine (T4) and calcitonin.

Thyroxine (T4) one of the hormones released from thyroid glands. It is involved in metabolism and development in the body.

Tinnitus ringing noise in the ear.

Tocolytic a drug, such as nifedipine and indomethacin, used to delay premature labour by reducing the contractions of the uterus.

Total lung capacity (TLC) the total lung volume or capacity is approximately 5–6 litres.

Tranexamic acid an inhibitor of fibrinolysis which is used to reduce excessive bleeding in menorrhagia.

Tricyclic antidepressants a group of anti-depressant drugs such as imipramine. They may exert their actions by inhibiting the uptake of some neurotransmitters, or may directly act on certain receptors. The actual mode of action is not known.

Triiodothyronine (T3) one of the hormones derived from tyrosine which is secreted by the thyroid gland.

Troponin protein involved in initiating the contraction process of skeletal and cardiac muscles. A rise in plasma concentration of troponin I and T is specific for cardiac injury.

Uraemia an increase in blood concentration of urea, usually due to malfunctioning of the kidney: a sign of kidney failure.

Valproate drug used for patients with epilepsy and also used in bipolar psychiatric disorders. It reduces or prevents epileptic seizures (anticonvulsant) and has the ability to stabilize mood in patients with bipolar disorder.

Valsartan an angiotensin 2 receptor antagonist.

Vasopressin ADH or arginine vasopressin, an antidiuretic hormone responsible for concentrating urine when the body is dehydrated.

Warfarin a drug with an anticoagulant activity. It prevents the formation of blood clot.

Index

Note that the numbers in this index refer to the Case Studies

Clinical Physiology and Pharmacology Farideh Javid and Janice McCurrie